SpringerBriefs in Earth System Sciences

SpringerBriefs South America and the Southern Hemisphere

Series Editors

Kevin Hamilton
Gerrit Lohmann
Lawrence A. Mysak
Jorge Rabassa

For further volumes:
http://www.springer.com/series/10032

Nat Rutter · Andrea Coronato
Karin Helmens · Jorge Rabassa
Marcelo Zárate

Glaciations in North and South America from the Miocene to the Last Glacial Maximum

Comparisons, Linkages and Uncertainties

 Springer

Nat Rutter
Department of Earth and Atmospheric
  Sciences
University of Alberta
Edmonton, AB T6G2E3
Canada

Andrea Coronato
Laboratorio de Geomorfología
  y Cuaternario
CADIC-CONICET
9410 Ushuaia
Argentina

Karin Helmens
Department of Physical Geography
  and Quaternary Geology
Stockholm University
10691 Stockholm
Sweden

Jorge Rabassa
Laboratorio de Geomorfología
  y Cuaternario
CADIC-CONICET
Bernardo Houssay 200
9140 Ushuaia
Argentina

Marcelo Zárate
INCITAP (CONICET-UNLPAM)
Avenida Uruguay 151
6300 Santa Rosa, La Pampa
Argentina

ISSN 2191-589X   ISSN 2191-5903 (electronic)
ISBN 978-94-007-4398-4   ISBN 978-94-007-4399-1 (eBook)
DOI 10.1007/978-94-007-4399-1
Springer Dordrecht Heidelberg New York London

Library of Congress Control Number: 2012939639

Printed on acid-free paper

Springer is part of Springer Science+Business Media (www.springer.com)

Acknowledgments

Nat Rutter acknowledges with thanks, constructive conversations with Alejandra Duk-Rodkin (Geological Survey of Canada) and Rene Barendregt (University of Lethbridge, Canada) on the glaciations of northwest North America. Thanks go to Igor Jakab (University of Alberta) for preparing Figs. 3.1, 3.2, 3.4, and 5.1, and Göran Alm and Krister Jansson (Stockholm University) for assistance in preparing Figs. 1.1 and 2.1. Jorge Rabassa and Andrea Coronato give thanks to students and colleagues who collaborated in field work in Patagonia and Tierra del Fuego, over 30 years of glacial geology research. Patagonian research has been supported by several grants from CONICET and ANPCyT (Argentina). Jorge Rabassa wants to express his deepest gratitude to Francisco Fidalgo, John Mercer and Calvin J. Heusser, who in the early 1970s introduced him in the wonderful world of Patagonian glaciations, and to Donald R. Coates for his unforgettable guidance and strong training in glacial geomorphology. The Bert Bolin Centre for Climate Research (BBCC) at Stockholm University and the University of Alberta Distinguished University Professor fund are acknowledged for financial aid in supporting two workshops (in Ushuaia, Argentina, and Stockholm, Sweden) for critical discussions pertinent to this paper.

Contents

Abstract

Improved dating methods have increased our ability to more precisely determine the timing and durations of glaciations. Utilizing glacial and loess deposits, we have compared glaciations that occurred in North and South America in order to determine whether events are synchronous or not, to explore forcing mechanisms, and to compare glaciations with cold periods of the Marine Oxygen Isotope stages and the loess/paleosol records of China. Stratigraphic sections containing a variety of glacial deposits, some with interbedded volcanics, as well as loess deposits, were used in reconstructing the glacial history. The major problems included fragmentary sections, missing evidence, and limited detailed age estimates. Dating methods utilized included radiocarbon, K–Ar, ^{40}Ar–^{39}Ar, fission track, paleomagnetic polarity interpretation, and luminescence techniques (China). In North America, the best evidence for Late Cenozoic glaciations is in southeastern Alaska, the Pacific Coastal Mountains, and northwestern Canada, in South America, in the Patagonian Andes and forelands of Argentina, and in the Bogotá basin of Colombia. Four Miocene-Pliocene (Gilbert and Gauss Chrons) and Pliocene–Pleistocene (Gauss/Matuyama Chrons) glaciations have been identified in southern South America, whereas only one mountain Pliocene and one Pliocene–Pleistocene glaciations have been identified in northwest America. This may be the result of the near-coeval volcanic activity which preserved till deposits under basaltic lava flows. Major tectonic uplift in the northern Andes suggests the absence of terrain high enough to support glaciers in the Bogotá mountains during the Miocene-Pliocene. During the Early and Late Pleistocene (Matuyama Chron) there are ten glacial advances recognized in either North or South America. Five appear to be roughly synchronous. During the Jaramillo Chron the greatest glaciation occurred in Patagonia whereas the Jaramillo glaciation in northwestern Canada was not a major event. At least three Middle Pleistocene (Brunhes Chron) glaciations are represented in both North and South America. In southern South America, subsequent glaciations are less extensive than the previous. The opposite is true in northern South America and North America where younger glaciations appear to be more extensive. This may indicate the impact of local factors such as tectonism, glacial overdeepining, and/or precipitation gradient increase acting over global forces. The Late Pleistocene (Brunhes Chron) Last Glacial Maximum is recognized in mountain and continental areas of North America but only in the mountains of South America. Commonly, our comparisons indicate roughly synchronous glaciations on the two continents, whereas other glaciations are more elusive and difficult to compare. Although our comparisons are at low resolutions, the results suggest that Milankovitch forcing is most likely the dominant trigger for hemispheric glaciation modified by local factors.

Introduction and Previous Work

Glacial events recorded on continents demonstrate that major climatic changes have taken place during much of Earth history. Correlation problems within regions, countries, and continents have centered around, among other things, few and questionable age estimates, incomplete knowledge of the extent of a glaciation, complicated by glacial deposits that may or may not represent a distinct advance, and glacial evidence showing similar characteristics among regions, countries, and continents but are not synchronous. In addition, variable tectonic uplift through time controlling the presence or absence of glaciers, principally along the western mountains of North and South America, where many of the best glacial sections are located, has hindered interpretation. Other problems include shifting climatic patterns affecting the presence or absence of glaciation. These problems have restricted correlation of glacial events between continents. By far the best understood is the later part of the last glacial cycle where moraines and stratigraphic sections are relatively well preserved and dating material is more abundant. One of the major efforts in Northern Hemispheric correlation was carried out by the International Geological Correlation Program (IGCP) 24 where scientists from many countries on both sides of the Atlantic investigated the possibility of synchronous (or non-synchronous) glacial events, mainly between Europe and North America (Sibrava and Richmond 1986). Since the IGCP 24 program ended, numerous studies have been published on glacial correlation but these are mainly confined to the Northern Hemisphere. In the Southern Hemisphere, only New Zealand and South America have extensive glacial records (Clapperton 1993; Suggate 1990). Apart from correlation for the last glacial cycle along the American Cordillera (Clapperton 2000), there has been little attempt for north–south correlation and comparisons.

The objective here is to compare and correlate late Tertiary and Pleistocene (i.e. from about 7 Ma to the Last Glacial Maximum (LGM) at 25–20 ka B.P.) glacial events between North and South America, focusing on the southern (Patagonian) Andes and adjacent Argentine plains; the northern Andes of Colombia; the North American plains and western mountains and the Arctic islands (Fig. 1).

The Patagonian Andes and forelands of Argentina provide a long record of glacial events. The sequences consist mostly of till, commonly separated by volcanic rocks, or lying on glacially eroded surfaces. The Bogotá basin in the Colombian Andes contains a long record of glacio-fluvial sediments interbedded with lacustrine sediments and tephras. In North America, glacial records are interpreted from sequences of till and stratified glacial sediments and erosional surfaces. In South America absolute age estimates are determined largely by ^{14}C, cosmogenic isotope exposure, ^{40}Ar–^{39}Ar, K–Ar and fission track radiometric methods from Andean volcanics associated with glacial and glacio-fluvial deposits or erosional surfaces. In North America radiocarbon ages for the last glacial cycle are abundant but ages for older deposits are rare. The age determinations in North America rely heavily on paleomagnetic, and other relative age dating methods, plus limited radiometric dates.

We discuss the major loess areas in North and South America with reference to the relatively complete Chinese loess/paleosol record. These records aid in defining cold periods that may or may not represent periods of glaciations but commonly support our glacial correlations. In addition, we discuss the glacial records according to the global oxygen isotope record of Lisiecki and Raymo (2005) and consider the base of the Quaternary at 2.6 Ma as recently approved by the International Union of Geological Sciences (IUGS).

Comparisons of chronology and extent of glaciations continue to be made on a worldwide scale but not in great detail (see Ehlers and Gibbard 2003, 2004a–c, 2007; Ehlers et al. 2011).

To the authors' knowledge, no comprehensive comparisons or correlation of glaciations have yet been made between North and South America for the entire Late Cenozoic sequence.

References

Clapperton C (1993) Quaternary geology and geomorphology of South America. Elsevier, Amsterdam, p 779

Clapperton C (2000) Interhemispheric synchroneity of marine oxygen isotope 2 glacier fluctuations along the American cordilleras transect. J Quat Sci 15:435–468

Ehlers J, Gibbard PL (2003) Extent and chronology of glaciations. Quat Sci Rev 22:1561–1568

Ehlers J, Gibbard PL (2004a) Quaternary glaciations—extent and chronology, part I: Europe. Developments in Quaternary Science, vol 2a, Elsevier, Amsterdam

Ehlers J, Gibbard PL (2004b) Quaternary glaciations—extent and chronology, part II: North America. Developments in Quaternary Science, vol 2b, Elsevier, Amsterdam

Ehlers J, Gibbard PL (2004c) Quaternary glaciations—extent and chronology, part III: South America, Asia, Africa, Australasia, Antarctica. Developments in Quaternary Science, vol 2c, Elsevier, Amsterdam

Ehlers J, Gibbard PL (2007) The extent and chronology of Cenozoic global glaciation. Quat Int 164–165:6–20

Ehlers J, Gibbard PL, Hughes PD (2011) Quaternary glaciations—extent and chronology. A closer look. Developments in Quaternary Science, vol 15, Elsevier, Amsterdam

Lisiecki LE, Raymo ME (2005) A Pliocene–Pleistocene stack of 57 globally distributed benthic $\delta^{18}O$ records. Paleoceanography 20(PA 1003):1–17

Sibrava VDQ, Richmond GM (eds) (1986) Quaternary glaciations in the northern hemisphere V. Quaternary science reviews, vol 5, Pergamon Press, Oxford, p 510

Suggate R (1990) Late Pliocene and Quaternary glaciations of New Zealand. Quat Sci Rev 9:175–197

Chapter 1
The Glacial and Loess Record
of Southern South America

Abstract The geomorphological, sedimentary and pedological evidence related to the various glacial periods which took place in the Patagonian and Fuegian Andes between the Late Miocene and the Late Pleistocene are herein presented and discussed. The occurrence of basaltic flows interbedded with till and other glaciogenic deposits provides radiometric dating and paleomagnetic information for several Patagonian regions which may be applied to the related glacial episodes. The modern glacial chronology of the Patagonian terrestrial glacial record is basically supported by $^{40}Ar/^{39}Ar$ dating on volcanic rocks associated with glacial deposits, and cosmogenic isotope dating techniques on erratic boulders and glacial erosion surfaces. These sequences, chronologically calibrated, compared with 47 Marine ^{18}O Isotope Stages (MIS) and constrained by global magnetostratigraphy, enable the interpretation of climatic changes over about seven million years in the Southern Hemisphere. In addition, the loess and loess-like record of southern South American extends back at least to the Late Miocene. These deposits cover the Chaco-Pampean region of central eastern Argentina, between 28–38° S, which formed a large continental sedimentary basin during the Neogene. The oldest glacial evidence is composed of till and glaciofluvial deposits interbedded with 5-7 Ma old basalts found in at least two areas, 38° and 46° S. It has been interpreted that the Pliocene–Pleistocene glaciations were of local ice cap and piedmont glacier type. These ancient glacial deposits have been also affected by Andean tectonics and deformation. At least parts of the Southern Andes were already covered by a local mountain ice sheet during the Late Pliocene. This statement is supported by the finding of glacial boulders in between basaltic flows at 41° S, lava flows overlying glacial and glaciofluvial units at 39° S, and glacial deposits interbedded with flows and till overlying basalt flows between latitude 46°–52° S. Whether the development of glaciers was promoted by climate deterioration, by tectonic forces, or both kinds of processes is still unknown. Thus, already during the Pliocene, local icefields would have been very extensive, covering part of the westernmost, extra-Andean tablelands all along Patagonia.

N. Rutter et al., *Glaciations in North and South America from the Miocene to the Last Glacial Maximum*, SpringerBriefs in Earth System Sciences, DOI: 10.1007/978-94-007-4399-1_1, © The Author(s) 2012

A minimum of seven glaciations were recognized at Cerro del Fraile locality (50° S) during the Early Pleistocene and interpreted as repeated piedmont glaciations, synchronous to active volcanism. The most extensive glaciation recognized in many parts of Patagonia is known as The Great Patagonian Glaciation (GPG) and bracketed between 1.15–1.05 Ma. However, until precise chronology is available for all localities in which GPG has been found, GPG should be considered as a multiple-event glaciation. In several transversal Patagonian valleys three glacial advances took place after GPG and before Last Glacial Maximum (LGM). The absolute chronologies of these post-GPG events have not been closely defined yet though, based upon paleomagnetic data, all till deposits along the Atlantic coast of Tierra del Fuego would belong to the Brunhes epoch. The Last Glaciation is very well known in all Patagonian glacial valleys and coastal channels and fjords. The estimated ages of these deposits begin close to 60 ka, but full glacial times (LGM) have been dated between 25–20 ka. Correlation of all these glacial events with the Pampean loess units is possible, but more detailed studies and absolute dating is needed.

Introduction

The Andean Cordillera extents from lat. 11°N–55° S along the western side of South America, with maximum elevations varying from 7,000–5,000 m in the Central and Northern Andes to 3,500–1,200 m in the Southern and Fuegian Andes (Fig. 1.1).

At present, the Andean Cordillera is affected by a wide range of climatic systems, from equatorial to subantarctic. Tectonism and global atmospheric changes during the past seven million years or so have promoted glaciation. Mountain glaciers developed in the high northern and central Andes, whereas large ice sheets produced outlet and piedmont glaciers along the southern Patagonian and Fuegian Andes.

In the central region of Argentina, late Miocene continental sedimentation is thought to have been triggered by a major pulse of tectonic uplift of the Andes, the Quechua phase circa 12 Ma (Marshall et al. 1983). As a result, predominantly clastic material including silt, sands and loess began to accumulate in the Pampas (Marshall et al. 1983). These deposits are regarded as sinorogenic sediments derived from the erosion of the uplifted Cordillera (Ramos 1999a) and distributed across a vast region, the Chaco-Pampean plain east of the Andes.

More recent studies, however, point to a diachronous uplift related to differences in the tectonic regime along the Andean Cordillera (Ramos 1999b). The uplift of the Patagonian Andes between 42°–47°S began around 20–17 Ma. Northward, the Cordillera Principal and Cordillera Frontal segments extending from 28°–38° S were elevated between 18–9 Ma and 16–3 Ma respectively. Therefore, from a glacial perspective, the diachronous process across the

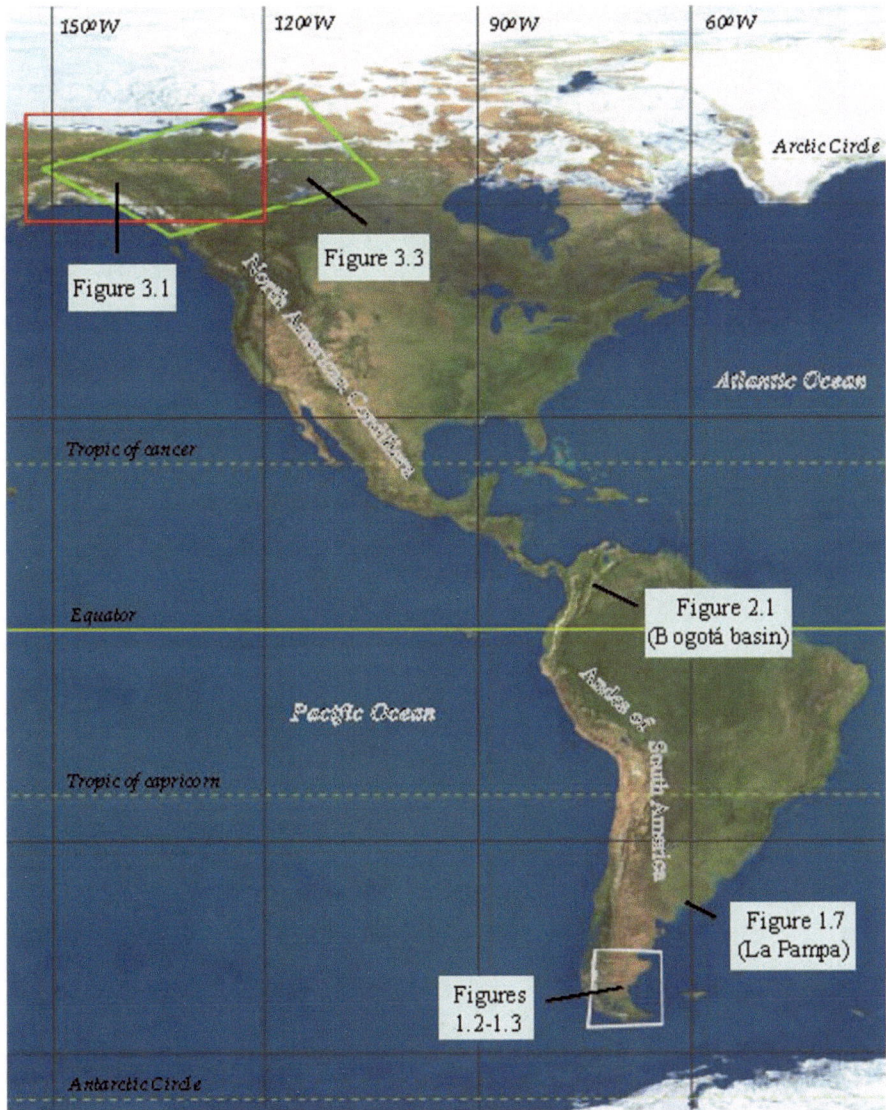

Fig. 1.1 Index map of North and South America indicating the position of various location maps used in the text

Cordillera must have generated the appropriate conditions for glaciation to develop at different time intervals at least since the Miocene.

The Patagonian glacial record is considered the best available in the Southern Hemisphere, as well as the oldest glacial record in the world outside Antarctica,

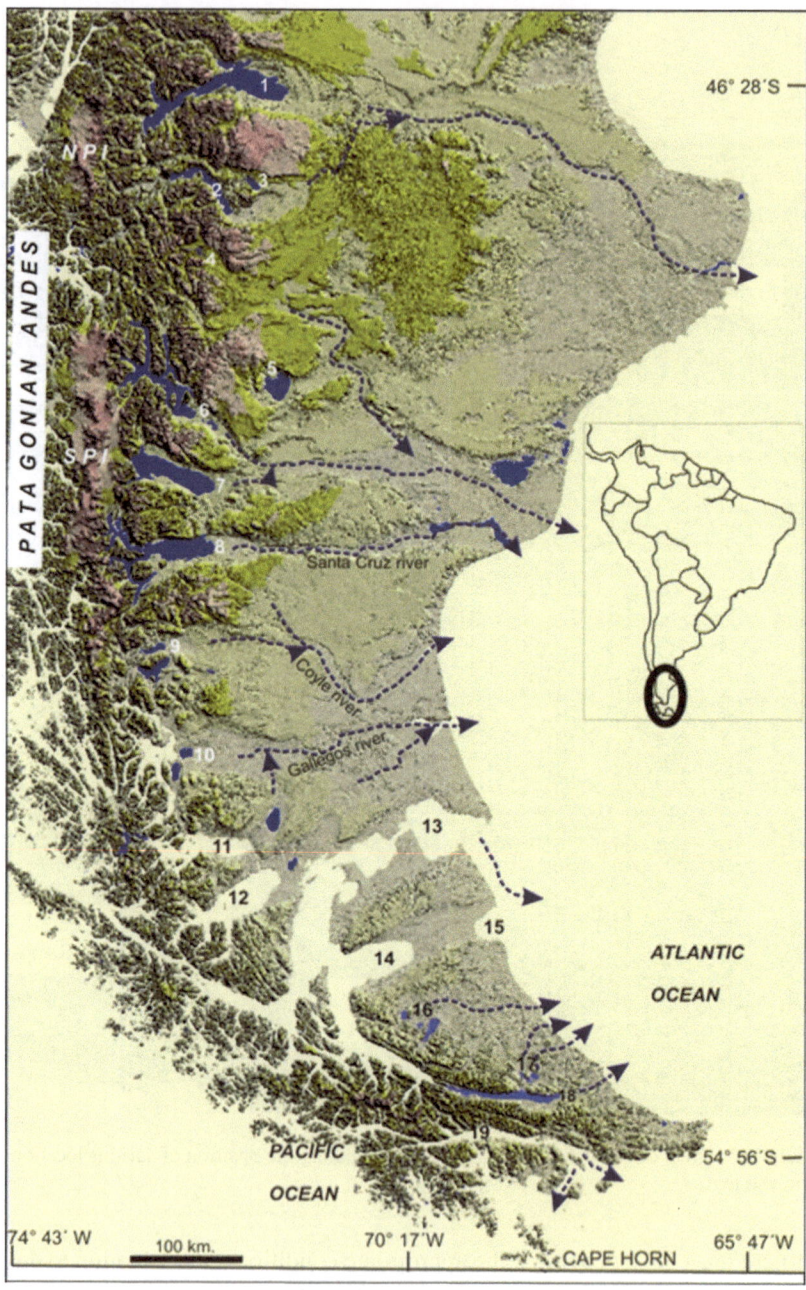

beginning from about 7–5 Ma–12 ka ago (Rabassa 1999; see Rabassa 2008, for a recent review, and all references cited therein; Figs. 1.2 and 1.3). Wenzens (2006) has proposed a glacier expansion as far east as Lago Cardiel (48° 54′S/

◄ **Fig. 1.2** Southern Patagonia and Tierra del Fuego location map. *Arrows* indicate palaeostreams flowing from Pleistocene glacial limits. The numbers indicate the name of the most important glacial valleys, as follows, *1* Lago Buenos Aires, *2* Lagos Pueyrredón-Posadas, *3* Lago Ghío, *4* Lagos Belgrano and Burmeister, *5* Lago Cardiel, *6* Lagos San Martín and Tarn, *7* Lago Viedma, *8* Lago Argentino, *9* Lagos Sarmiento and Toro, *10* Lagos Balmaceda and Pinto, *11* Sound Skyrring, *12* Sound Otway, *13* Straits of Magellan, *14* Bahía Inútil, *15* Bahía San Sebastián, *16* Lagos Lynch and Blanco, *17* Lagos Yehuin and Chepelmut, *18* Lago Fagnano, *19* Beagle Channel (modified from Coronato and Rabassa 2011)

71° 13'W) during the Middle Miocene, ca. 10–11 Ma, but the true glacial nature of the exposed units there has been challenged by experts in Patagonian regional geology (R. de Barrio, Universidad de La Plata and J.L. Panza, Argentine Geological Survey, personal communication to J.R.). Further work is still needed to confirm the glacial origin of the cited sediments. The modern glacial chronology of the Patagonian terrestrial glacial record is based upon ^{40}Ar/^{39}Ar dating on volcanic rocks associated with glacial deposits, and cosmogenic isotope dating techniques on erratic boulders and glacial erosion surfaces. These sequences, chronologically calibrated, compared with the marine ^{18}O Isotope Stages (MIS) and constrained by global magnetostratigraphy, enable the interpretation of climatic changes over about seven million years in the Southern Hemisphere. In addition, the loess and loess-like record of southern South American extends back at least to the late Miocene (Marshall et al. 1983; Zárate 2003). These deposits cover the Chaco-Pampean region of central eastern Argentina, between 28°–38°S, which formed a large continental sedimentary basin during the Neogene.

The correlation of the loess and loess-like record with the Andean glacial record has not been attempted until recently (Rabassa et al. 2005), because of poor chronological control and the nature of the sedimentary sequences. With the exception of numerical ages mostly obtained for loess-paleosol sequences of the last glacial cycle (among others, Kemp et al. 2004a, b, 2006), the chronology of deposits older than 100 ka have been primarily based on relative ages inferred from fossil mammal assemblages (South American land mammal ages or SALMA, sensu Pascual et al. 1965) and magnetostratigraphy (Valencio and Orgeira 1983; Ruocco 1989; Orgeira 1987, 1990). More recently numerical ages on impact glasses obtained at key localities of the southern Pampas together with new magnetostratigraphic analysis (Schultz et al. 1998, 2004, 2006) have provided a tool to better calibrate some of the Late Miocene and Pliocene land mammal ages. The sedimentary sequences are dominantly composed of reworked loess (loess-like deposits) re-sedimented by aqueous transport agents; a high degree of bioturbation, and pedogenesis, including abundant carbonate nodules and calcareous cementation and crusts, resulting in complex stratigraphic successions. The primary eolian origin of the sediments is revealed by their allochtonous nature (Andean derived volcanic rock particles) of the loess-like deposits that blanket Precambrian and Paleozoic-Mesozoic bedrock of central Argentina.

The Glacial Record

Glaciations During the Latest Miocene, Early Pliocene and Middle Pliocene (Gilbert Chron)

Along the Río Aluminé Valley (38° 57'S/71° 02'W), northern Patagonia, Schlieder (1989) recognized conglomerates interpreted as distal glaciofluvial units of the Andean glaciations. Their age is limited by overlying basalts (6.41 ± 0.13 and 5.26 ± 0.14 Ma), thus major glaciations occurred in northern Patagonia during the latest Miocene, as it had been suggested by Gracia (1958) and González Díaz and Nullo (1980).

The Meseta del Lago Buenos Aires (Fig. 1.2) contains till deposits interbedded with basaltic flows (Mercer 1976; Clapperton 1993). Mercer (1976) and Mercer and Sutter (1981) obtained whole-rock, K/Ar ages for the under- and overlying lavas of 7.34 ± 0.11–6.75 ± 0.08 and 5.05 ± 0.07–4.43 ± 0.09 Ma, respectively, indicating a latest Miocene age for the till (Ardolino et al. 1999). In this locality, Thon-That et al. (1999) obtained $^{40}Ar/^{39}Ar$, incremental heating ages of 7.38 ± 0.05 and 5.04 ± 0.04 Ma for the underlying and overlying flows, roughly confirming Mercer and Sutter (1981) ages; thus, ice caps would have extended close to the Gilbert and C3 magnetic chron boundary. Limiting ages for the tills suggest that major extra-Andean glaciations could have taken place during the late Gilbert Chron, including the Sidujfall and Thevra Subchrons. Although more than one till may be involved in these ancient deposits, the glacial event is called, in general terms as the Lago Buenos Aires I Glaciation.

This area, including the neighboring Meseta del Guenguel, has been also studied in detail for the time-space relationshipos between tectonics, volcanism and glaciation (Lagabrielle et al. 2004, 2007, 2010). These authors have also established the presence of Late Miocene-Pliocene glaciation in the Meseta Lago Buenos Aires, certainly between 7 and 3 Ma ago, but perhaps even as piedmont glaciation as old as ca. 10 Ma. The observed deposits suggest that a local ice cap and piedmont glaciers existed in this area in Pre-Quaternary times. These ancient glacial deposits have been also affected by Andean tectonics and deformation.

Glaciations During the Late Pliocene (Gauss Chron)

In northern Patagonia, Rabassa et al. (2005, 2011) inferred that at least part of the Andes were already covered by a local mountain ice sheet during the Gauss-Gilbert Chron boundary. This statement was postulated by finding glacial boulders in basaltic flows at Mount Tronador (41° 10'S/ 71° 53'W) dated by K/Ar at 3.2 and 2.0 Ma, and lava flows overlying glacial and glacial fluvial deposits in the Pulmarí, Quillén and Alicura valleys (39° 10'S/ 70° 55'W), $^{39}Ar/^{40}Ar$ (whole rock) dated at 3.5–3.4 Ma (Schlieder 1989; Rabassa et al. 2011). However, these units should be redated with more modern techniques and, until then, it is not possible to confirm glacial stages during this period in Northern Patagonia.

In southern Patagonia, in the Lago Viedma region (Fig. 1.2), Mercer (1976) and Mercer et al. (1975) identified glacial deposits interbedded with flows, K/Ar dated at 3.55 ± 0.19 and 3.68 ± 0.03 Ma; another till overlies a flow dated at 3.46 ± 0.22 Ma. A till layer occurs between flows dated at 3.48 ± 0.09 and 3.55 ± 0.07 Ma, in the early Gauss Chron. Tills overlie and underlie lava flows dated at 3.20 Ma (Lago Argentino) and 3.45 Ma (Lago Viedma) (Mercer 1976). The older till is named for the Lago Viedma Glaciation whereas the younger till represents the Lago Argentino Glaciation. These dates correspond to the Mammoth subchron (3.3 Ma). At Cóndor Cliff (lat. 50°S), along the Santa Cruz valley flow (Fig. 1.2), burying till was dated by Mercer (1976) at 2.79 ± 0.15 Ma. Thus, glaciation took place during the late Gauss Chron, showing that even during the Pliocene some glaciers reached locations close to the maximum Pleistocene expansion (the Great Patagonian Glaciation, GPG). This may represent a separate glacial event that followed the Lago Argentino Glaciation and is tentatively named as the Río Santa Cruz Glaciation. Whether the development of glaciers was promoted by climate deterioration, by tectonic forces or both is not known yet. Thus, already during the Pliocene, icefields would have been very extensive, covering part of the extra-Andean tablelands all along Patagonia.

Recently, an outstanding, relatively complete sediment core has been retrieved from the Ross Embayment of Antarctica (McKay et al. 2009). Three pre-Pleistocene glacial advances have been recognized there. These are a "polar type" advance that took place during the late Middle Miocene, and a "subpolar type" advance that took place during the Late Miocene. These fall into the time limits of the Late Pleistocene glaciations in the Patagonian Andes, Argentina. A later "subpolar" to "polar" advance was identified as being formed during the Pliocene-Early Pleistocene and may be equivalent to the Lago Viedma II or later glaciations. Figure 1.3 shows the extent of these ancient glaciations in the southern Andes in violet and yellow colors.

Glaciations During the Early Pleistocene (Matuyama Chron)

Wenzens (2000) obtained limiting ages of 3.0–2.25 Ma for glacigenic deposits at Lago Viedma, indicating that glaciation may have occurred along the Late Gauss-Early Matuyama Chrons boundary. Mörner and Sylwan (1989) and Sylwan (1989) mentioned till at Lago Buenos Aires during the Early Matuyama Chron, in coincidence with Wenzens (2000, 2006) and Schellmann (1998, 1999). These deposits then represent the Lago Buenos Aires Glaciation.

Glacigenic sequences at Cerro del Fraile (Figs. 1.3 and 1.4) interbedded with flows, were firstly considered to be of "Pliocene" age (Feruglio 1944), even before the time of radiometric dating, and when the Pleistocene epoch was restricted to the last million years of Earth's history. Later, these flows were K/Ar dated

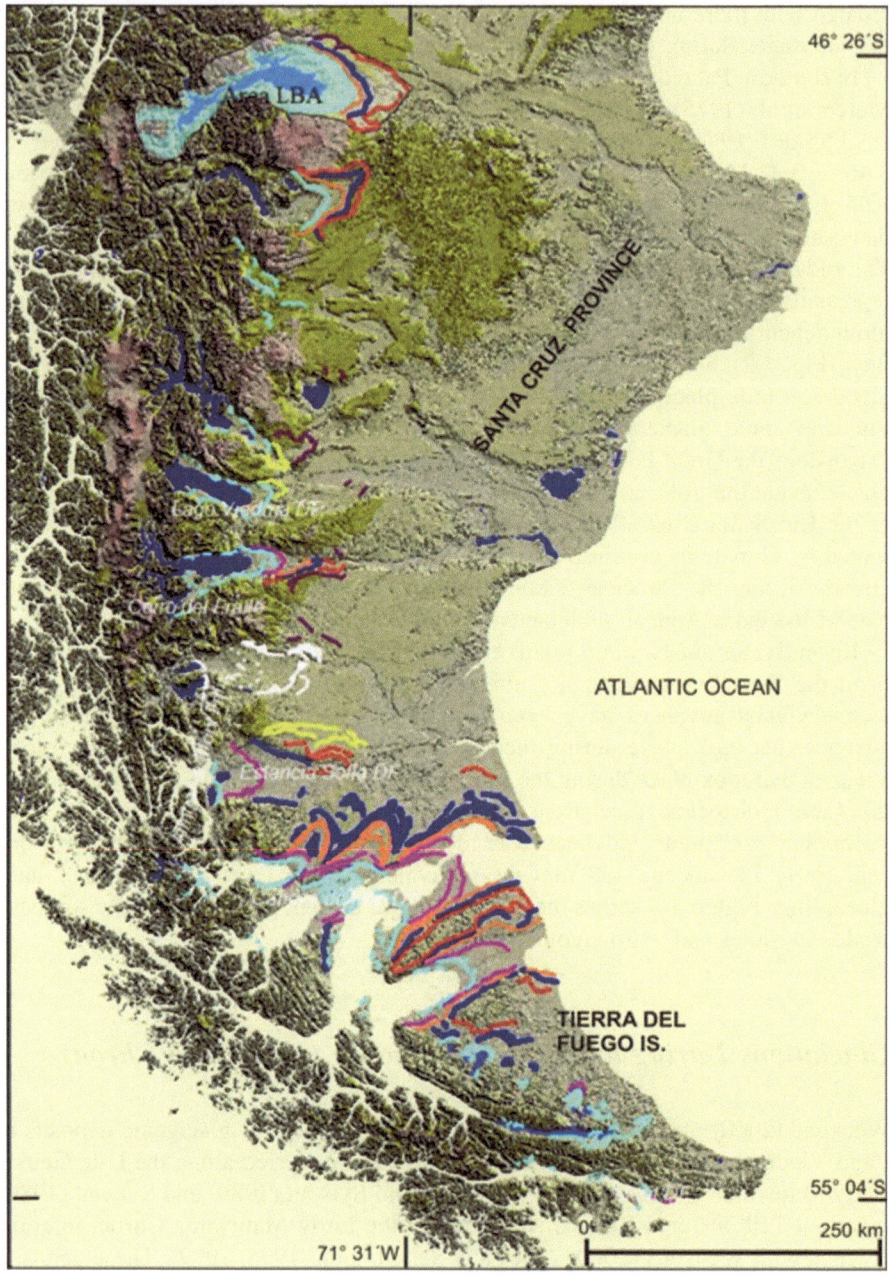

Fig. 1.3 Extent of Early-Middle Pleistocene glaciations along the main glacial valleys of the Southern Patagonian and Fuegian Andes. For name of each glacial valley see Fig. 1.2. The estimated age for each glacial limit is indicated by colors, as follows, *violet* Pre-Great Patagonian Glaciation (*Pre-GPG*); *yellow* Pre-GPG of unknown age; *red* GPG; *blue* Post-GPG 1; *orange* Post-GPG 2; *pink* Post-GPG 3; *light blue* Last Glacial Maximum (*LGM*); *dashed light blue* Late Glacial. Modified from Coronato and Rabassa (2011)

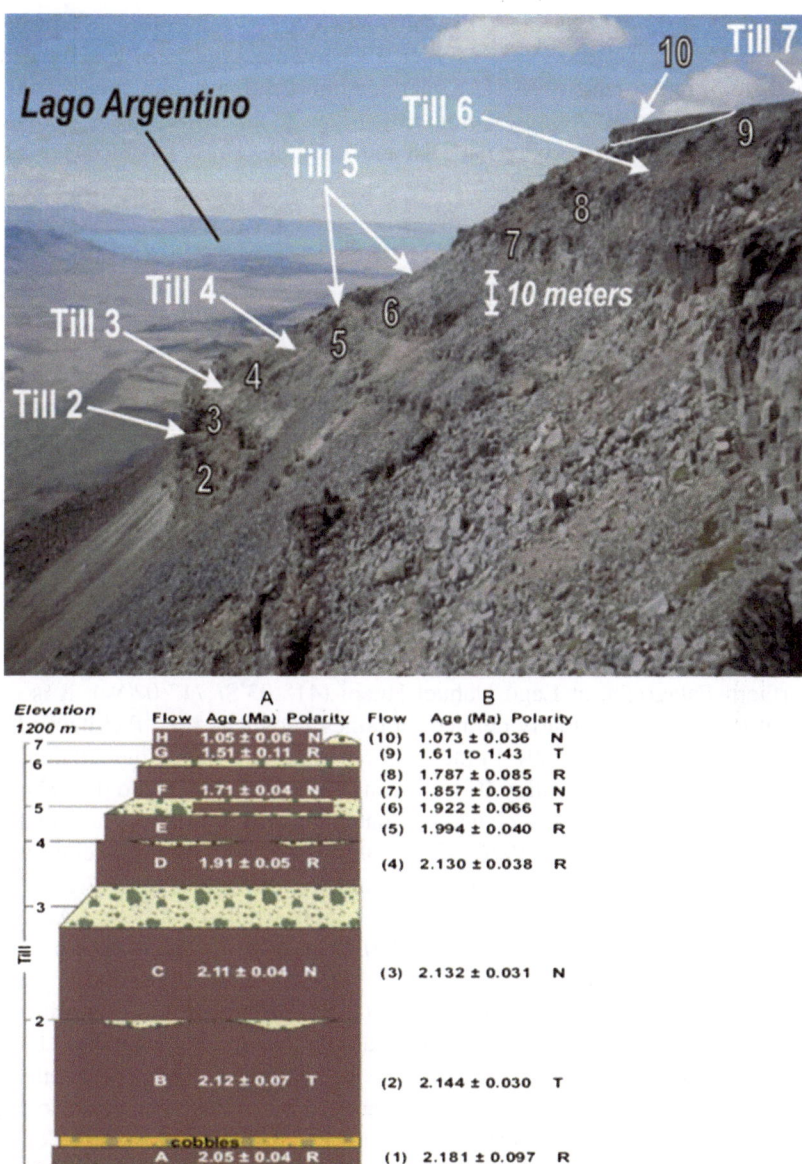

Fig. 1.4 Cerro del Fraile locality (1,200 m a.s.l.), southern Patagonia; from Singer et al. (2004b). Inter-bedded till layers deposited between lava flows show that seven glacial advances affected the extra-andean Patagonian high plateau reliefs between 2.1–1.0 My. Magnetic polarity studies were performed by (*A*) Fleck et al. (1972) and (*B*) Singer et al. (2004b)

between 2.08–1.03 Ma, during the Matuyama Chron, suggesting six piedmont glaciations (Mercer et al. 1975; Mercer 1976). This sequence was redated by magnetostratigraphy and ^{40}Ar/^{39}Ar incremental heating techniques (Ton-That et al. 1999; Singer et al. 2004b). A minimum of seven glaciations were recognized, which most likely occurred between 2.18 and 1.43 Ma. These advances are named as the Cerro del Fraile Glaciations I to IV.

The Great Patagonian Glaciation

The most extensive glaciation has been recognized in many parts of Patagonia (Fig. 1.3). The Great Patagonian Glaciation (GPG; Mercer 1976; Rabassa and Clapperton 1990; Coronato et al. 2004a; Singer et al. 2004a) represents the maximum expansion of the ice from the Andes. Its geographical distribution was correctly mapped by Caldenius (1932) and corresponds to his "Initioglacial" stage, though wrongly interpreted as an early phase of the last glaciation or to the penultimate glaciation. Feruglio (1950) already suggested that each one of the morainic systems recognized by Caldenius (1932) was in fact representing a different glaciation, being the older one of Early Pleistocene age. The extent of glaciations in each glacial valley for southernmost Patagonia is shown in Fig. 1.3. In Northern Patagonia, at Lago Nahuel Huapi (41° 03'S/ 71° 02'W), it is represented at least by three clearly defined moraines known as the "Pichileufú Drift" (Flint and Fidalgo 1964, 1969). In the southern end, ice reached the Atlantic coast south of the Río Gallegos valley (Figs. 1.2 and 1.3), expanding into the submarine platform. The age of this advance was estimated by various methods and authors in southern Patagonia. Mercer (1976) K/Ar dated flows underlying glacial deposits south of the Río Gallegos Valley obtaining ages between 1.47 ± 0.1 and 1.17 ± 0.05 Ma. Meglioli (1992) used total fusion, whole-rock ^{40}Ar/^{39}Ar method obtaining basal limiting ages of 1.55 ± 0.03 Ma at the Bella Vista Basalt (Río Gallegos Valley) which was overridden by glaciers and is covered by erratics. Ton-That et al. (1999) redated this basalt (^{40}Ar–^{39}Ar incremental heating technique) at 1.168 ± 0.007 Ma. These authors also provided a reliable upper limit for the GPG by means of ^{40}Ar/^{39}Ar dating of the Telken Basalt, which covers the GPG deposits at Lago Buenos Aires (Fig. 1.3), at 1.016 ± 0.005 Ma. See also Singer et al. (2004a). At Cerro del Fraile, a till covers the uppermost flow, dated at 1.08 Ma (Guillou and Singer 1997; Singer et al. 2004b) being probably equivalent to the GPG (Fig. 1.4). It is then possible to place the GPG during the Late Matuyama Epoch, close to the Jaramillo normal Sub-chron. Anyway, until precise chronology is available, the GPG should be considered as a multiple-event glaciation. Similar conclusions have been recently achieved by the study of multiple, superposed tills with sand-wedge (Fig. 1.5) features and their stratigraphical relationships, near Río Gallegos (Bockheim et al. 2009).

In southern Patagonia, in the Lago Buenos Aires region (Fig. 1.3), Ton-That (1997) and Ton-That et al. (1999) obtained limiting ages by ^{40}Ar/^{39}Ar (incremental

Fig. 1.5 Two sets of sand wedge casts penetrating till at Tres de Enero locality, southern Patagonia, are shown. Till of this locality is correlated to the Sierra de los Frailes drift, deposited during the Great Patagonian Glaciation (GPG). The exposure shows two till units both affected by periglacial features, the development of four cold periods, two as glacial advances and two as tundra environments. This evidence allows interpreting the so-called Great Patagonian Glaciation as a multiple-event glaciation. For sand wedge cast details see Bockheim et al. (2009)

Fig. 1.6 Middle Pleistocene glacial deposits and erratic boulders at Punta Sinaí, Tierra del Fuego. **a** Cliff erosion along the Southern Atlantic coast in Tierra del Fuego displays ancient glacial deposits. **b** Surface morphology shows an undulated moraine with kettle-holes and high erratic boulders being exhumed by erosion. Although palaeomagnetic studies gave a Brunhes Chron for the glacial deposits (Walther et al. 2007), the cosmogenic exposure age could not be established for this drift unit (Kaplan et al. 2004), which has been proposed as Post-Great Patagonian Glaciation 1 by Coronato et al. (2004b). **c** and **d** Basal till exposure showing boulders of various sizes at Punta Sinaí cliff

heating technique) dating the Arroyo Page Basalt (0.760 ± 0.007 Ma), which covers recessional outwash deposits that appear to represent the Daniglacial or Post GPG I Glaciation that took place following the GPG.

Glaciations During the Middle-Late Pleistocene (Brunhes Chron)

The most widespread glacial event at the end of the Middle Pleistocene took place during the Gotiglacial period (Caldenius 1932) called the Post GPG III Glaciation (Coronato et al. 2004a, b). However, in southernmost localities of the region, a previous glaciation, called the Post-GPG II Glaciation, has been recognized (Coronato et al. 2004a, b). The Gotiglacial deposits show very well-preserved moraines, located above the last glaciation units and several tens of kilometers downvalley from the younger terminal moraines. In southern Patagonia, in Lago Buenos Aires, a flow from Cerro Volcán (normal magnetic polarity) postdates the post-GPG II and post-GPG III (Gotiglacial) deposits and predates those of the last glaciation (Coronato et al. 2004a). This flow was dated (whole-rock K/Ar) by Mercer and Sutter (1981) in 0.177 ± 0.056 Ma. Later, Ton-That et al. (1999) obtained a ^{40}Ar/^{39}Ar plateau age of 0.123 ± 0.005 Ma and an unspiked K/Ar age of 0.128 ± 0.002 Ma (Guillou and Singer 1997). These dates were later confirmed by cosmogenic isotope exposure dates, which provided a mean age of 0.128 ± 0.003 Ma (Ackert et al. 1998; Singer et al. 1998). It is thus possible to confirm a minimum age of late Middle Pleistocene for the Gotiglacial advances (post-GPG III, post-GPG II), but the precise age of these deposits must be further investigated. Southwards in the Lago Pueyrredón valley (47° 30'S, Fig. 1.3), ^{10}Be cosmogenic surface exposure dating performed on cobbles from old outwash terraces gave a mean age of 0.260 Ma (Hein et al. 2009) which could give a maximum age for the post GPG III glaciation.

In Tierra del Fuego, a preliminary paleomagnetic study at Punta Sinaí and San Sebastián, south of the Bahía San Sebastián–Bahía Inútil depression, provided normal polarity data for glacial deposits there, thus assigning them a Brunhes age (Walther et al. 2007). These units were interpreted as Post GPG II and Post GPG III (Coronato et al. 2004b). Figure 1.3 shows the limits of Middle Pleistocene glaciations along the southern Patagonian Andes (Fig. 1.6).

In the Central Andes, at Cerro Aconcagua (32° 39'S/ 70°W), the Uspallata and Punta de Vacas drifts have been identified (Espizúa 1993, 2004), as of Middle Pleistocene age. Although the chronology is not yet confirmed, the oldest glaciation is thought to be early Middle Pleistocene and the youngest would be ca. 260–134 ka (Espizúa and Bigazzi 1998).

The Late Pleistocene (Last Glacial Cycle: Brunhes Chron)

The glacigenic deposits of the last glaciation are those formed in the Late Pleistocene, after the Sangamon interglacial. The estimated age of the deposits of this period begin close to 60 ka, but glacial deposits from the full glacial age (LGM) have been dated between 25–20 ka. In several Patagonian regions, glacial advances predating the LGM have been identified (see Coronato and Rabassa 2007). Using TL and U-Series methods and ^{14}C dating, these advances have been constrained between 48 and 35–28 ka. This means that cold conditions affected all of South America, after the Sangamon and before the LGM, though perhaps differentially according to latitude.

In the Aconcagua valley, the Penitentes, Horcones, Almacenes and Confluencia drifts have been recognized by Espizúa (1993). While the Penitentes Drift represents an ice advance about 40 ka B.P., the Horcones Drift stands for full glacial conditions at 21 ka B.P., whereas the Almacenes and Confluencia drifts correspond to Late Glacial and Neoglacial times. This glacial model is questioned today due to the work of Fauqué et al. (2009) that demonstrates that debris flow and mega landslides, including glacier boulders, have occurred several times during the mid and late Pleistocene affecting younger glacial deposits, re-shaping them as mass wasting landforms at the bottom of the Horcones and Cuevas valleys. Based upon ^{36}Cl cosmogenic dated boulders and radiocarbon dates, the latter authors argue that mass wasting processes had a stronger impact on landscape development than mid-late Pleistocene glaciers.

It is clear that it is still poorly known how glaciers reacted to Late Pleistocene cooling in the arid Andes. Palaeoclimatic, sedimentological and modelling data suggest a more humid climate than today, allowing snowline lowering. An LGM stage has been interpreted between 30–20 ka B.P. in northern Chile. LGM was recognized also in the Aconcagua massif, where end moraines older than 15 ka (TL data) are found in valleys at 2,700 m elevation about 22 km down valley from the cirques (Coronato and Rabassa 2007).

Along the Patagonian Andes, LGM was named as "Finiglacial" by Caldenius (1932), "Nahuel Huapi drift" by Flint and Fidalgo (1964) and "Llanquihue drift" by Clapperton (1993). The more reliable chronological dates in the North Patagonian Andes come from Lago Llanquihue (41° 14'S/ 73° 03'W). There, studies by Mercer (1976), Porter (1981) and Lowell et al. (1995) provided an adjusted chronology based on ^{14}C dates. According to these authors, ice expanded in the early Late Pleistocene (Early Wisconsinan), receded afterwards (Mid Wisconsin Interstadial; Laugenie 1984; Rabassa and Clapperton 1990) and further readvanced in the latest Late Pleistocene (Late Wisconsinan). The earlier advances of the glaciers were generally reached and more commonly even surpassed by later readvances.

Kaplan et al. (2004) confirmed the LGM age through cosmogenic isotope dating, differentiating five glacial episodes at Lago Buenos Aires. The respective ages extend from 25 ka for the outermost Fénix V to 16 ka for the innermost Fénix I moraines. An AMS ^{14}C age of 15.3 ± 0.3 ka BP in post-LGM glaciolacustrine deposits validates these exposure ages and provides an upper limiting age for LGM in this region. Further south, in the Magellan Straits area (Figs. 1.2 and 1.3), McCulloch et al. (2005a) identified five glacial stages. Based on ^{14}C and isotopic analyses, they proposed that LGM occurred after 31.2 cal ka B.P. (Stage A) and culminated at ca. 25.2–23.1 cal ka B.P. (Stage B); a less extensive advance occurred later (Stage C) and the glacier retreated before 21.7–20.4 cal ka B.P. The third major advance (Stage D) occurred sometime before 17.5–16.6 ka B.P. The chronology of the last ice advance is not clear. The evidence indicates that it would have occurred between 15.5–14.3 or between 12.5–11.7 cal ka B.P. (McCulloch et al. 2005b), during the Antarctic Cold Reversal (ACR), that is, the early Late Glacial (Late Pleistocene) climate deterioration shown by the Vostok ice-core paleotemperature record (Blunier and Brook 2001).

On the Isla Grande de Tierra del Fuego, glacial morphology located upslope into the cirques has been mostly interpreted as belonging to the LGM (Rabassa et al. 2000). By contrast to the Late Glacial, neither radiocarbon or cosmogenic isotope chronologies confirms the LGM chronology. Lateral morainic deposits located along the Río Fuego valley, TL-dated at 25.7 ka (Coronato et al. 2008) is correlated with Stage B in the Bahía Inútil lobe following McCulloch et al. (2005a).

In summary, then, the identified glaciations in southern South America are listed in Table 1.1.

Table 1.1

Period	Chron	Southern South American glaciation	Estimated age (Ma)	Probable MIS
L. Miocene/				
E. Pliocene	Gilbert	Lago Buenos Aires	7.00–4.400	>T2
L. Pliocene	Gauss	Lago Viedma I	3.45–3.35	MG6–MG8
L. Pliocene	Gauss	Lago Argentino	3.30–3.25	M2
L. Pliocene/	Gauss/			
/E. Pleistocene	/Matuyama	Lago Viedma II	3.00–2.350	92, 96, 98, 100, 104, G6, G10
E. Pleistocene	E. Matuyama	Cerro del Fraile I	<2.180 to >2.6	90, 98, 100, 104
E. Pleistocene	E. Matuyama	Cerro del Fraile II	<2.144 to >2.132	78, 82
E. Pleistocene	E. Matuyama	Cerro del Fraile III	<2.132 to >2.130	78
E. Pleistocene	E. Matuyama	Cerro del Fraile IV	<2.130 to >1.99	74, 76
E. Pleistocene	L.Matuyama/Olduvai	Cerro del Fraile V	<1.99 to >1.85	70, 72
E. Pleistocene	L. Matuyama	Cerro del Fraile VI	<1.78 to >1.43	50, 52, 54, 58
E. Pleistocene	L. Matuyama/Jaramillo	Great Patagonian Glaciation	1.15–1.05	30
E. Pleistocene	L. Matuyama	Post GPG I	<1.10 to >0.76	20, 22
M. Pleistocene	Brunhes	Post GPG II	<0.76	16, 18
M. Pleistocene	Brunhes	Post GPG III	0.26–0.1	6, 8
L. Pleistocene	Brunhes	Last Glacial Maximum (LGM)	0.048–0.025	2, 4

The Loess Record

Miocene-Pliocene Loess (Gauss-Gilbert Chron)

Late Miocene loess and loess-like deposits have been reported from several localities of southern Buenos Aires and La Pampa provinces, Argentina (Figs. 1.1 and 1.7). The sequence comprises a 150–200 m thick sedimentary apron covering both tectonic basins and uplands. The successions are composed of high percentages of volcanic lithic particles (andesitic and basaltic rocks) exceeding 50 %. Several paleosol levels are present suggesting an episodic sedimentation under semiarid to subhumid environmental conditions (Visconti 2007). The late Miocene sedimentation process and the formation of soils is roughly bracketed between <9 and 5.3 Ma by fossil mammal assemblages (Huayquerian South American land mammal age SALMA; Montalvo et al. 2007, and references therein) and numerical ages on impact glasses (Schultz et al. 2006) obtained from sedimentary units which are stratigraphically underlying and overlying the succession.

The late Miocene and Pliocene pattern of alternation of paleosols and loess/loess-like deposits suggest some sort of cyclicity in the sedimentation process. Kemp and Zárate (2000) performed a detailed micromorphological analysis at a selected Pliocene section of the Chapadmalal locality inferring pedosedimentary cycles and a balance between sedimentation and pedogenesis. Based on this, it was speculated that the inferred cycles, including intervals of higher sedimentation rates of loess might be related to Pliocene glacial advances that caused an increased transport of sediment from the high Andes by meltwater streams into floodplains. In turn, wind deflated and transported the sediments on to the southern Pampas (Kemp and Zárate 2000). As records of late Miocene and Pliocene glaciations have been reported from several different areas of the Northern Patagonian Andes of Neuquén (Rabassa et al. 2011) it seems now more plausible to hypothesize that the cyclicity observed in the Pampean successions during this time interval is the result of changing climatic conditions. In summary, tectonism must have generated the conditions for glaciation to develop which controlled the climatic conditions. A Pliocene loess record bearing Chapadmalalan SALMA fossil remains (Marshall et al. 1983) is exposed south of Mar del Plata, along the Chapadmalal sea-cliff (Fig. 1.7). The succession is composed of Andean volcaniclastic siltstones and clay siltstones regarded as loess deposits (Teruggi 1957; Zárate 2003) comprising seven major paleosols that occur at regular stratigraphic intervals throughout the exposed succession and are laterally traceable for several kilometers, suggesting major intervals of landscape stability. A Gilbert age was attributed to the succession (Orgeira 1990) which, together with numerical dates obtained on an upper layer of impact glasses and magnestostratigraphic analysis, suggests that this loess-paleosol record encompasses approximately the time interval between 5–4.5 and 3.3 Ma (Schultz et al. 1998).

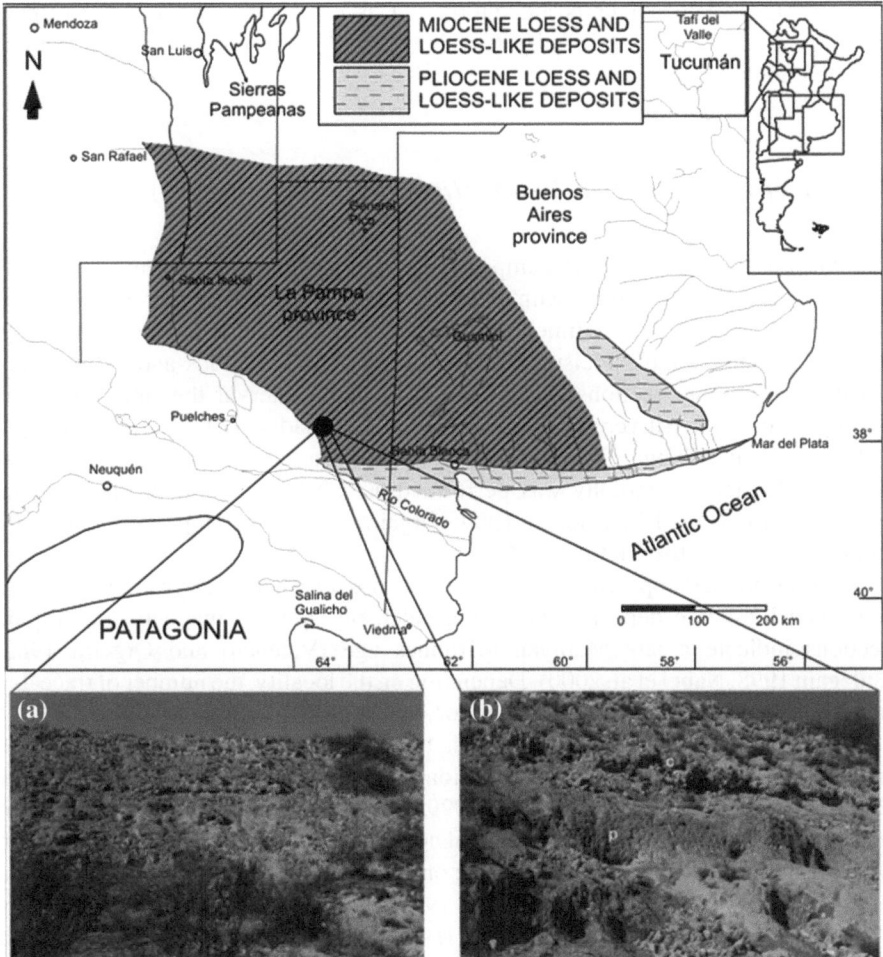

Fig. 1.7 Areal distribution of Late Miocene and Pliocene loess and loess-like deposits mostly covered by a thin mantle of Late Quaternary aeolian sediments. **a** General view of a Late Miocene loess-paleosol succession in the SW of La Pampa Province; **b** detail of paleosol (*p*) overlain by a carbonate layer (*c*)

A later succession caused by excavation of fluvial valleys initiated the sedimentation of fluvial channel facies followed by the accumulation of loess deposits including at least two well-developed paleosols which are laterally traceable along several kilometers. Fossil mammal assemblages of the Marplatan SALMA (Cione and Tonni 1995) were recovered from this stratigraphic interval. On the basis of stratigraphic relationships and magnestostratigrahy suggesting a late Gilbert to early Gauss age (Orgeira 1990), loess accumulation may have occurred sometimes between 3.3 and 2.6 Ma.

These loess-paleosol units are interpreted as cold and warm periods respectively, with the two loess units correlated to the pre-Pleistocene glaciations reported above.

Pleistocene Loess (Brunhes-Matuyama-Gauss Chrons)

The Pleistocene record of the Pampean region, particularly the Matuyama stratigraphic interval, is mostly documented by loess-like deposits representing both fluvial and paludal-like sedimentary environments, which include interbedded paleosols. The available Pleistocene outcrops are discontinuous and expose only partial stratigraphic sections. Within the Matuyama interval the time control is mainly based on fossil vertebrate assemblages (Ensenadan SALMA, Pascual et al. 1965; Cione and Tonni 1995) and do not provide an adequate chronological resolution. Magnetostratigraphy was performed at two localities along the Mar del Plata sea-cliff (Fig. 1.7; Ruocco 1989; Orgeira 1987), encompassing relatively minor intervals of the Matuyama Chron.

In the northern Pampas, the exposed successions up to nearly 18 m thick consist of loess and loess-like deposits with some paleosols. Several magnetostratigraphic sections indicate a late Matuyama-Bruhnes age (Valencio and Orgeira 1983; Bidegain 1998; Nabel et al. 2000). Depending on the locality, the number of traceable and discrete paleosols varies between 3 or 4 throughout this time interval.

Northwestward of the Pampean region, the 40–50 m thick mountain loess record of Tafí del Valle, Tucumán, at an elevation of 2,300 m (Fig. 1.6) includes between 28–32 paleosols (Zinck and Sayago 1999, 2001; Schellenberger and Veit 2006) which suggest cyclicity of environmental conditions. The sequence was interpreted as recording part of the last glacial cycle on the basis of ^{14}C dates with a maximum age of 40 ka B.P. (Zinck and Sayago 1999, 2001). Later, luminescence dating yielded ages of circa 190 ka at the lower levels; see Kemp et al. 2003. A much older age of around 1 Ma was inferred from magnetostratigraphy interpreting the succession as an almost continuous and complete record of climatic changes and environmental conditions of subtropical South America (Schellenberger et al. 2003; Schellenberger and Veit 2006).

Last Glacial Cycle (Brunhes Chron)

Luminescence dating, micromorphology and sedimentology of several localities across the Pampean plain (Kemp et al. 2004b, 2006) indicate the predominance of loess deposition during the last glacial cycle. The Late Pleistocene (Early and Mid Wisconsinan) time interval is mostly recorded by reworked eolian facies (loess-like deposits). In the western part of the northern Pampas (province of Córdoba, Fig. 1.6), loess accumulation seems to have continued until the mid-Holocene when soil development began (Kemp et al. 2006). Also, a diachronous paleosol probably spanning the equivalent of at least part of the Sangamonan Interglaciation has been identified and traced across several localities of the region.

The Pleistocene record of central Argentina, mostly composed of loess-like deposits, suggests that Andean volcanic sediments from the northernmost Patagonian Andes (lat. 38°–40°S) and perhaps the central Andes of Mendoza and San Juan (lat. 32°–38°S) were transported to the Pampean plain. At late Matuyama-Brunhes exposures of the northern Pampas, the occurrence of several paleosols suggests cyclicity, with periods of higher sedimentation rates and intervals with dominance of pedogenesis. Although a pattern of loess-paleosol cycles was proposed (e.g. Tonni et al. 1999) detailed analysis at a representative section of the regional stratigraphy indicates a changing balance of sedimentation, pedogenesis and erosion through this time interval that does not conform to the classical loess-paleosol model (Zárate et al. 2002).

In the early Pleistocene, the available Pampean sections are of low stratigraphic resolution which along with the lack of detailed studies, constrain any adjusted correlations with the Andean glaciations. In turn, assuming that paleosols represent interglacial intervals, the record of the late Matuyama-Brunhes age of the northern Pampas, encompassing around 1 Ma, includes only 3–4 regional paleosols and therefore does not reflect the global climatic pattern. This is either because of the dominant sedimentary conditions in the northern Pampas that gave way to paleosol welding, low stratigraphic sequences of low resolution, or the way global climatic changes sedimentation rates and continuous pedogenesis resulting in more condensed stratigraphic sequences of low resolution, or the way global climatic changes manifested in this part of the Southern Hemisphere. Only the last glacial loess-paleosol sequence seems to document global climatic conditions.

References

Ackert RP, Singer B, Guillou H, Kurz M (1998) Cosmogenic ^{3}He production rates over the last 125,000 years: calibration against $^{40}Ar/^{39}Ar$ and unspiked K-Ar ages of lava flows. Geological Society of America 1998 annual meeting symposium 24, Abstracts, Toronto

Ardolino A, Franchi M, Remersal M, Salani F (1999) El volcanismo en la patagonia extraandina. In: Haller M (ed) Geología Argentina. Anales SEGEMAR N° 29, Buenos Aires, pp 579–612

Bidegain JC (1998) New evidence of the Brunhes-Matuyama polarity boundary in the Hernández-Gorina quarries, North West of the city of La Plata, Buenos Aires Province, Argentina. Quat S Am Antarct Peninsula 11:207–228

Blunier T, Brook E (2001) Timing of millenial-scale climate change in Antarctica and Greenland during the last glacial period. Science 291:109–112

Bockheim J, Coronato A, Ponce JF, Ercolano B, Rabassa J (2009) Relict sand wedges in southern Patagonia and their stratigraphic and paleoenvironmental significance. Quatern Sci Rev 28(13–14):1188–1199

Caldenius C (1932) Las glaciaciones cuaternarias en Patagonia and Tierra del Fuego. Anales, Dirección General de Geología y Minería. Buenos Aires 95, p 150

Cione A, Tonni EP (1995) Chronostratigraphy and "land-mammal ages" in the Cenozoic of southern South America: principles, practices, and the "Uquian" problem. J Paleontol 69:135–159

Clapperton C (1993) Quaternary geology and geomorphology of South America. Elsevier, Amsterdam, p 779

Coronato AMJ, Rabassa J (2007) Late Quaternary glaciations in South America. In: Scott E (ed) Encyclopedia of Quaternary Science, vol 2. Elsevier, Amsterdam, pp 1101–1108

Coronato AMJ, Martínez O, Rabassa J (2004a) Glaciations in Argentine Patagonia, southern South America. In: Ehlers J, Gibbard JP (eds) Quaternary glaciations: extent and chronology. Part III, South America, Asia, Africa, Australia, Antarctica. Developments in Quaternary Science, vol 2c. Elsevier, Amsterdam, pp 49–67

Coronato AMJ, Meglioli A, Rabassa J (2004b) Glaciations in the Magellan Straits and Tierra del Fuego, southernmost South America. In: Ehlers J, Gibbard JP (eds) Quaternary glaciations: extent and chronology. Part III: South America, Asia, Africa, Australia, Antarctica, Developments in Quaternary Science, vol 2c. Elsevier, Amsterdam, pp 45–48

Coronato AMJ, Ponce JF, Seppälä M, Rabassa J (2008) Englazamiento del valle del Río Fuego durante el Pleistoceno tardío, Tierra del Fuego, Argentina. Actas del XVII Congreso Geológico Argentino, pp 1194–1195

Coronato AMJ, Rabassa J (2011) Pleistocene glaciations in Southern Patagonia and Tierra del Fuego. In: Ehlers J et al. (eds) Quaternary glaciations: extent and chronology. A closer look. Developments in Quaternary Science, Chap. 51, vol 15. Elsevier, Amsterdam

Espizúa L (1993) Quaternary glaciations in the Río Mendoza valley, Argentine Andes. Quatern Res 40:150–162

Espizúa L (2004) Pleistocene glaciations in the Mendoza Andes, Argentina. In: Ehlers J, Gibbard P (eds) Quaternary glaciations-extent and chronology, Part III: South America, Asia, Africa, Australasia, Antarctica. Developments in Quaternary Science, vol 2c. Elsevier, Amsterdam, pp. 69–73

Espizúa L, Bigazzi G (1998) Fission-track dating of the Punta de Vacas Glaciation in the Río Mendoza valley, Argentina. Quatern Sci Rev 17:755–760

Fauqué L, Hermanns R, Hewitt K, Rosas M, Wilson C, Baumann V, Lagorio S, Di Tomasso I (2009) Mega-deslizamientos de la pared sur del Cerro Aconcagua y su relación con depósitos asignados a la glaciación pleistocena. Revista de la Asociación Geológica Argentina 65(4):691–712

Feruglio E (1944) Estudios geológicos y glaciológicos en la región del Lago Argentino (Patagonia). Bol Acad Nac Ciencias Córdoba 37:1–208

Feruglio E (1950) Descripción geológica de la Patagonia. T. 3, Cuaternario. Y.P.F., Buenos Aires

Fleck RJ, Mercer JH, Nairn AEM, Peterson DN (1972) Chronology of Late Pliocene and Early Pleistocene glacial and magnetic events in southern Argentina. Earth Planet Sci Lett 16:15–22

Flint RF, Fidalgo F (1964) Glacial geology of the east flank of the Argentine andes between latitude 39°10′S and latitude 41°21′S. Geol Soc Am Bull 75:335–352

Flint RF, Fidalgo F (1969) Glacial drift in the eastern Argentine andes between latitude 41°10′S and latitude 43°21′S. Geol Soc Am Bull 80:1043–1052

González Díaz EF, Nullo F (1980) Cordillera Neuquina. In: Leanza A (ed) Geología Regional Argentina, vol 2. Academia Nacional de Ciencias de Córdoba, Argentina, pp 1099–1148

Gracia, R (1958) Informe geológico de las cartas Paso Flores y Traful. Secretaría de Ejército, Dirección General de Ingenieros, Buenos Aires, unpublished report

Guillou H, Singer B (1997) Combined unspiked K-Ar and ^{40}Ar/^{39}Ar dating of Late Quaternary lavas. EOS, Transactions of the American Geophysical Union, Abstracts, Fall Meeting 78, 46, 771

Hein A, Hulton N, Dunai T, Schnabel C, Kaplan M, Naylor M, Xu S (2009) Middle Pleistocene glaciations in Patagonia dated by cosmogenic-nuclide measurements on gravels. Earth Planetary Sci Lett 286:184–197

Kaplan MR, Ackert RP Jr, Singer BS, Douglass DC, Kurz MD (2004) Cosmogenic nuclide chronology of millenial-scale glacial advances during O-isotope stage 2 in Patagonia. Geol Soc Am Bull 116:308–321

Kemp R, Zárate M (2000) Pliocene sedimentary cycles in the southern Pampas Argentina. Sedimentology 47:481–488

Kemp R, Toms P, Sayago JM, Derbyshire E, King M, Wagoner L (2003) Micromorphology and OSL dating of the basal part of the loess-paleosol sequence at La Mesada in Tucumán province, Northwest Argentina. Quaternary Int 106–107:111–117

Kemp R, King M, Toms P, Derbyshire E, Sayago JM, Collantes M (2004a) Pedosedimentary development of part of a Late Quaternary loess-paleosol sequence in northwest Argentina. J Quat Sci 19:1–10

Kemp RA, Toms PS, King M, Kröhling DM (2004b) The pedosedimentary evolution and chronology of Tortugas, a Late Quaternary type-site of the northern Pampa, Argentina. Quatern Int 114:101–112

Kemp R, Zárate M, Toms P, King M, Sanabria J, Arguello G (2006) Late Quaternary paleosols, stratigraphy and landscape evolution in the northern Pampas, Argentina. Quatern Res 66:119–132

Lagabrielle Y, Suárez M, Rossello E, Hérail G, Martinod J, Régnier M, De La Cruz R (2004) Neogene to Quaternary tectonic evolution of the Patagonian Andes at the latitude of the Chile triple junction. Tectonophysics 385:211–241

Lagabrielle Y, Suárez M, Malavieille J, Morata D, Espinoza F, Maury RC, Scalabrino B, Barbero L, de la Cruz R, Rossello E, Bellon H (2007) Pliocene extensional tectonics in the eastern central Patagonian Cordillera: geochronological constraints and new field evidence. Terra Nova 19:1–12. doi:10.1111/j.1365-3121.2007.00766.x

Lagabrielle Y, Scalabrino B, Suárez M, Ritz JF (2010) Mio-Pliocene glaciations of central Patagonia: new evidence and tectonic implications. Andean Geol 37(2):276–299

Laugenie C (1984) Le dernier cycle glaciaire quaternaire et la construction des nappes fluviatiles d'avant pays dans les Andes Chiliennes. Bulletin Association Française d'études de Quaternaire 1–3:139–145

Lowell T, Heusser CJ, Andersen B, Moreno P, Hauser A, Heusser L, Schlüchter C, Marchant D, Denton G (1995) Interhemispheric correlation of Late Pleistocene glacial events. Science 269:1541–1549

Marshall L, Hoffstetter R, Pascual R (1983) Mammals and stratigraphy: geochronology of the continental mammal-bearing tertiary of South America. Palaeovertebrata, Mémoire Extraordinaire, pp 1–93

McCulloch R, Fogwill C, Sudgen D, Bentley M, Kubik P (2005a) Chronology of the last glaciation in central Straits of Magellan and Bahía Inútil, southernmost South America. Geografiska Annaler, 87 A, 2, pp 289–312

McCulloch R, Bentley M, Tipping R, Clapperton C (2005b) Evidence for late-glacial ice dammed lakes in the central Straits of Magellan and Bahía Inútil, southernmost South America. Geografiska Annaler, 87 A, 2, pp 335–362

McKay R, Browne G, Carter L, Dundar G, Krissek L, Naish T, Powell R, Reed J, Talarica F, Wilch T (2009) The stratigraphic signature of the late Cenozoic Antarctic ice sheets in the Ross Embayment. Geol Soc Am Bull 121:1537–1561

Meglioli A (1992) Glacial geology of southernmost Patagonia, the Strait of Magellan and Northern Tierra del Fuego. Unpublished Ph.D. Dissertation, Lehigh University, Bethlehem, Pennsylvania, USA

Mercer JH (1976) Glacial history of southernmost South America. Quatern Res 6:125–166

Mercer J, Sutter J (1981) Late Miocene-earliest Pliocene glaciation in southern Argentina: implications for global ice-sheet history. Palaeogeogr Palaeoclimatol Palaeoecol 38:185–206

Mercer JH, Fleck RJ, Mankinen EA, Sander W (1975) Southern patagonia: glacial events between 4 and 1 MY ago. In: Suggate RP, Cresswell MM (eds) Quaternary studies. Royal Society of New Zealand Bulletin 13, pp 223–230

Montalvo CI, Melchor R, Visconti G, Cerdeño E (2007) Vertebrate taphonomy in loess-palaeosol deposits: a case study from the Late Miocene of central Argentina. Geobios 41:133–143

Mörner N, Sylwan C (1989) Magnetostratigraphy of the Patagonian moraine sequence at Lago Buenos Aires. J S Am Earth Sci 2:385–390

Nabel PE, Cione A, Tonni E (2000) Environmental changes in the Pampean area of Argentina at the Matuyama-Brunhes boundary (C1r–C1n) chrons boundary. Palaeogeogr Palaeoclimatol Palaeoecol 162, pp 403–412

Orgeira MJ (1987) Estudio paleomagnético de sedimentos asignados al Cenozoico tardío aflorantes en la costa Atlántica bonaerense. Revista de la Asociación Geológica Argentina 42:362–376

Orgeira MJ (1990) Paleomagnetism of Late Cenozoic fossiliferous sediments from Barranca de los Lobos (Buenos Aires Province, Argentina). The magnetic age of South American land mammal ages. Phys Earth Planet Inter 64:121–132

Pascual R, Ortega Hinojosa J, Gondar D, Tonni EP (1965) Las edades del Cenozoico mamalífero de la Argentina, con especial atención a aquellas del territorio bonaerense. Anales Comisión de Investigaciones Científicas de la Provincia de Buenos Aires 6, La Plata, pp 165–193

Porter SC (1981) Pleistocene glaciation in the southern lake district of Chile. Quatern Res 16:263–292

Rabassa J (1999) Cuaternario de la cordillera Patagónica y Tierra del Fuego. In: Haller M (ed) Geología Argentina, vol 29. Anales SEGEMAR N°, Buenos Aires, pp 710–714

Rabassa J (2008) Late Cenozoic glaciations in Patagonia and Tierra del Fuego. In: Rabassa J (ed) Late Cenozoic of Patagonia and Tierra del Fuego, Developments in Quaternary Sciences, vol 11. Elsevier, Amsterdam, pp 151–204

Rabassa J, Clapperton CM (1990) Quaternary glaciations in the southern Andes. Quatern Sci Rev 9:153–174

Rabassa J, Coronato AMJ, Bujalesky G, Roig C, Salemme M, Meglioli A, Heusser CJ, Gordillo S, Roig Juñent F, Borromei A, Quattrocchio M (2000) Quaternary of Tierra del Fuego, southernmost South America: an up-dated review. Quatern Int 68–71:217–240

Rabassa J, Coronato AMJ, Salemme M (2005) Chronology of the Late Cenozoic Patagonian glaciations and their correlation with biostratigraphic units of the Pampean region (Argentina). J S Am Earth Sci 20:81–103

Rabassa J, Coronato AMJ, Ponce JF, Schlieder G, Martínez OA (2011) Depósitos glacigénicos (Cenozoico tardío-Cuaternario) y geoformas asociadas. Relatorio del XVIII Congreso Geológico Argentino, Neuquén, pp 295–314

Ramos V (1999a) Los depósitos sinorogénicos terciarios de la región andina. In: Haller M (ed) Geología Argentina, vol 29. Anales SEGEMAR N°, Buenos Aires, pp 651–682

Ramos V (1999b) Las provincias geológicas del territorio argentino. In: Haller M (ed) Geología Argentina, vol 29. Anales SEGEMAR N°, Buenos Aires, pp 41–96

Ruocco M (1989) A 3 Ma paleomagnetic record of coastal continental deposits in Argentina. Palaeoecol, Palaeogeogr, Palaeoclimatol 72:105–113

Schellenberger A, Veit H (2006) Pedostratigraphy and pedological and geochemical characterization of Las Carreras Loess. Quatern Sci Rev 25:811–831

Schellenberger A, Heller F, Veit H (2003) Magnetostratigraphy and magnetic susceptibility of Las Carreras loess-paleosol sequence in Valle de Tafí, Tucumán NW-Argentina. Quaternary Int 106(107):159–167

Schellmann G (1998) Jungkanozoische landschaftsgesschichte patagoniens (Argentinien). Andine vorlandvergletscherungen, talentwicklung und marine terrasen. Essener Geogr Arbeiten 29:1–218

Schellmann G (1999) Landscape evolution and glacial history of southern Patagonia (Argentina) since the Late Miocene—some general aspects. Zentralbatt Geologie und Palaontologie, Teil I 7/8:1013–1026

Schlieder G (1989) Glacial geology of the Northern Patagonian Andes between lakes Aluminé and Lácar. Unpublished Ph.D dissertation, Lehigh University, Bethlehem, Pennsylvania, USA

Schultz PH, Zárate MA, Hames W, Camilión C, King J (1998) A 3.3 Ma impact in Argentina and possible consequences. Science 282:2061–2063

Schultz PH, Zárate MA, Hames W, Koeberl C, Bunch T, Storzer D, Renne P, Wittke J (2004) The Quaternary impact record from the Pampas, Argentina. Earth Planet Sci Lett 219:221–238

Schultz PH, Zárate MA, Hames W, Harris S, Bunch TW, Koeberl RP, Wittke J (2006) The record of Miocene impacts in the Argentine pampas. Meteorit Planet Sci 41:749–771

Singer B, Ackert RP, Kurz M, Guillou H, Ton-That T (1998) Chronology of Pleistocene glaciations in Patagonia: a ^{3}He, ^{40}Ar/^{39}Ar and K-Ar study of lavas and moraines at Lago Buenos Aires, 46°S, Argentina. Geological Society of America 1998 Annual meeting, symposium 24, Abstracts 30, 299, Toronto

Singer B, Ackert RP, Guillou H (2004a) ^{40}Ar/^{39}Ar ages and K/Ar chronology of Pleistocene glaciations in Patagonia. Geol Soc Am Bull 116:434–450

Singer B, Brown LL, Rabassa J, Guillou H (2004b) ^{40}Ar/^{39}Ar ages of Late Pliocene and Early Pleistocene geomagnetic and glacial events in southern Argentina. In: Timescales of the internal geomagnetic field. Am Geophys Union Monogr Ser 145:175–190

Sylwan C (1989) Paleomagnetism, paleoclimate and chronology of Late Cenozoic deposits in southern Argentina. Meddelanden Stockholms Universitets Geologiska Instituut, 277, p 110

Teruggi ME (1957) The nature and origin of Argentine loess. J Sediment Petrol 27:322–332

Tonni EP, Nabel P, Cione AL, Etchichury M, Tófalo R, Scillato Yané G, San Cristóbal J, Carlini A, Vargas D (1999) The Ensenada and Buenos Aires formations (Pleistocene) in a quarry near La Plata, Argentina. J S Am Earth Sci 12:273–291

Ton-That T (1997) 40Ar/39Ar dating of basaltic lava flows and the geology of the Lago Buenos Aires region, Santa Cruz province, Argentina. Unpublished Diploma Thesis, Université de Genève, Switzerland, p 51

Ton-That T, Singer B, Mörner NA, Rabassa J (1999) Datación por el método ^{40}Ar/^{39}Ar de lavas basálticas y geología del Cenozoico superior en la región del Lago Buenos Aires, provincia de Santa Cruz, Argentina. Revista de la Asociación Geológica Argentina 54(4):333–352

Valencio DA, Orgeira MJ (1983) La magnetoestratigrafía del Ensenadense y Bonaerense de la ciudad de Buenos Aires: Parte II. Revista de la Asociación Geológica Argentina 38(1):24–33

Visconti G (2007). Sedimentología de la Formación Cerro Azul (Mioceno superior) de la provincia de La Pampa, Argentina. Unpublished Doctoral Thesis, Universidad de Buenos Aires, Buenos Aires, Argentina

Walther A, Rabassa J, Coronato AMJ, Tassone A, Vilas JF (2007) Paleomagnetic study of glacigenic sediments from Tierra del Fuego. Geosur, Abstract, vol 174

Wenzens G (2000) Pliocene piedmont glaciation in the Río Shehuen valley, southwest Patagonia, Argentina. Arct Antarct Alp Res 32:46–54

Wenzens G (2006) Terminal moraines, outwash plains, and lake terraces in the vicinity of Lago Cardiel (49° S; Patagonia, Argentina)—evidence for Miocene foreland glaciations. Arct Antarct Alp Res 38:276–291

Zárate MA (2003) The loess record of southern South America. Quatern Sci Rev 22:1987–2006

Zárate MA, Kemp RA, Blasi AM (2002) Identification and differentiation of Pleistocene paleosols in the northern pampas of Buenos Aires, Argentina. J S Am Earth Sci 15:303–313

Zinck JA, Sayago JM (1999) Loess-paleosol sequence of La Mesada in Tucumán province, northwest Argentina—characterization and palaeoenvironmental interpretation. J S Am Earth Sci 12:293–310

Zinck JA, Sayago JM (2001) Climatic periodicity during the Late Pleistocene from a loess-paleosol sequence in northwest Argentina. Quatern Int 78:11–16

Chapter 2
The Glacial Record of Northern South America

Abstract The Bogotá basin and direct surroundings (eastern Andes in Colombia) holds a long sedimentary sequence that reaches from the present into the Miocene. Palaeo-botanical data indicates tectonic uplift by some 2,000 m during the Late Miocene–Pliocene that probably precluded sufficiently high enough terrain to support glaciers during the time period preceding the Quaternary. The first mountain glaciation as recorded by glaciofluvial sedimentation in the Bogotá basin is dated by fission-track and magnetic polarity dating at ca. 2.6 Ma, whereas a shift towards more extensive glaciations occurred after ca. 0.8 Ma. Moraines preserved in the Bogotá mountains record a series of glacial events for the time interval ca. 43 to 12.5 ka BP. Equilibrium Line Altitude (ELA) depression by ca. 1,300 m is reconstructed for the early Last Glacial Maximum (LGM; ca. 20 ka BP).

Introduction

The Colombian Andes in tropical South America (between latitudes 1–11°N) rises to elevations just over 5,000 m and is presently glaciated only very locally (see Helmens 2004). The total area of glaciated terrain was substantially enlarged during the Pleistocene, estimated to roughly 7.5 % of the total surface area of the Andean mountains (Thouret et al. 1996). The Pleistocene glacial record has been studied in detail in the high plain of the Bogotá area in the eastern Andes (Fig. 1.1). The high plain of Bogotá is situated at an elevation of 2,600 m and represents the bed of a former lake that occupied the tectonic basin of Bogotá during the Late Pliocene–Pleistocene (Fig. 2.1).

The Bogotá area holds an exceptionally long continental sediment sequence that reaches from the present into the Miocene (Van der Hammen et al. 1973; Helmens and Van der Hammen 1994; Van der Hammen and Hooghiemstra 1997). The sequence is near continuous in the Bogotá basin which began to accumulate

N. Rutter et al., *Glaciations in North and South America from the Miocene to the Last Glacial Maximum*, SpringerBriefs in Earth System Sciences, DOI: 10.1007/978-94-007-4399-1_2, © The Author(s) 2012

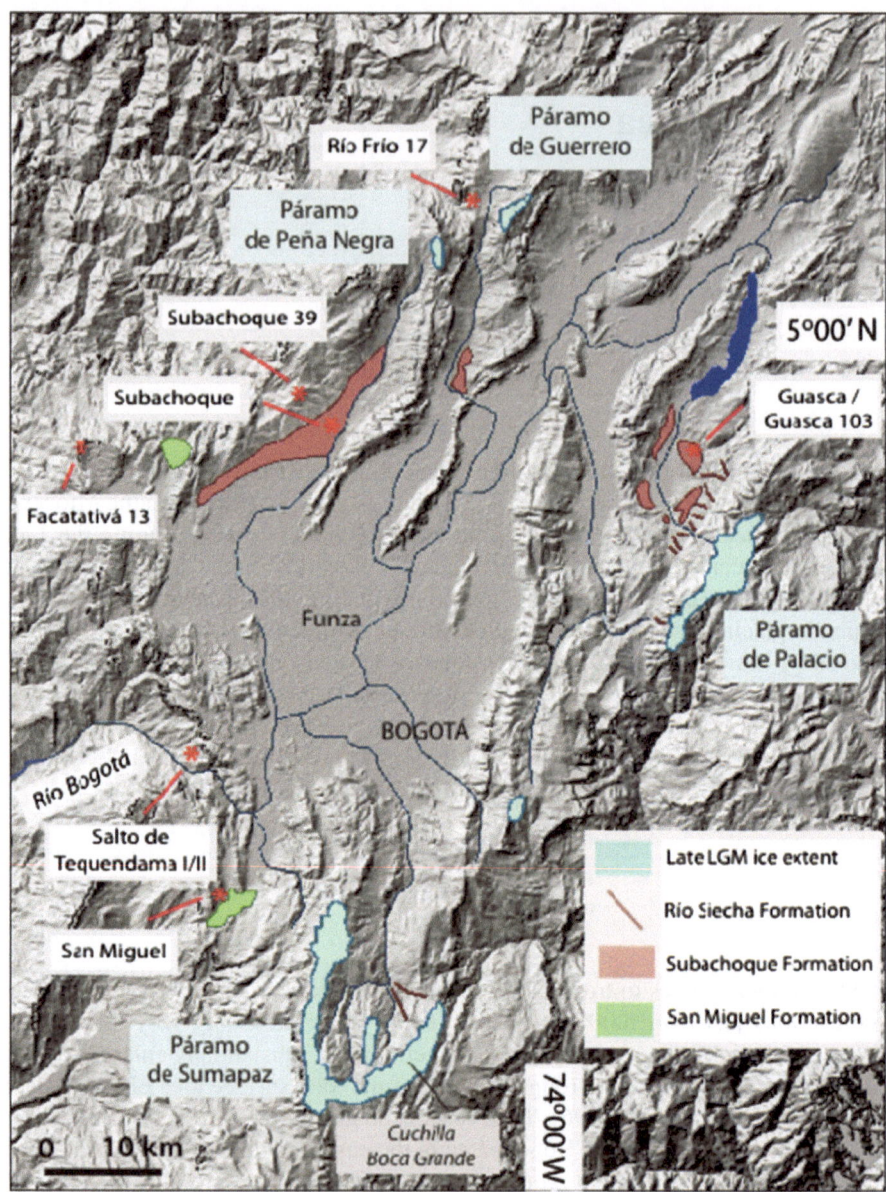

Fig. 2.1 The high plain of Bogotá and surrounding mountains (Eastern Colombian Andes; Fig. 1.1), showing the late LGM extent of mountain glaciers and the surface distribution of the Río Siecha (outwash fans), Subachoque (glaciofluvial and lake sediments) and San Miguel Formations (slope deposits). Location of exposures/boreholes mentioned in the text is also shown (based on Helmens 1990)

sediments at about 3 Ma. The older part of the sequence occurs as fragmented sections in the hills to the west of the basin. The sediments have been formally defined into 16 lithostratigraphic units and their surface distribution and geomorphic expression have been mapped over a total area of nearly 2,000 km^2 (Helmens 1990). Palynological data obtained from organic-bearing beds in exposed sections and from continuous lacustrine sequences from the central Bogotá basin, including the nearly 600 m long Funza II record (Hooghiemstra and Ran 1994), forms the basis for a biostratigraphic framework of 7 biozones (Van der Hammen et al. 1973; Kuhry and Helmens 1990; Wijninga 1996a). Absolute chronological control is provided through fission-track dating of volcanic ash layers (Helmens 1990; Andriessen et al. 1993; Helmens et al. 1997a), magnetic polarity dating (Helmens et al. 1997a), and radiocarbon dating of sediments and Andosols (summarized in Van der Hammen et al. 1980; Helmens 1990; Helmens and Kuhry 1995).

The palaeo-botanical record of the Bogotá area registers major tectonic uplift for the period between about 6–3 Ma. The uplift is recorded in the sections Salto de Tequendama I/II (Van der Hammen et al. 1973; Wijninga 1996b), Río Frío 17 (Wijninga 1996c), Subachoque 39 (Wijninga and Kuhry 1990), Facatativá 13 (Van der Hammen et al. 1973; Wijninga 1996d) and Guasca 103 (Wijninga and Kuhry 1993) (for locations see Fig. 2.1). Sediments of Middle Miocene age have provided fossil evidence of predominantly warm tropical lowland plant taxa, whereas Late Miocene and Pliocene sediments show increasing proportions of plant taxa associated with high-elevation (Biozones I-III). This trend in vegetation has been primarily interpreted in terms of a gradual tectonic uplift of the Eastern Andes in the Bogotá region by some 2,000 m during the Late Miocene-Pliocene (Van der Hammen et al. 1973; Kuhry and Helmens 1990; Wijninga 1996a). Figure 2.2 shows the inferred elevation at which the analyzed sediment sections were originally deposited, taking into account an estimated uncertainty of about 500 m. The latter corresponds to a range in temperature variation of about 3 °C expected for the Pliocene (Wijninga 1996a). Age control is based on fission-track dating of intercalated tephras (sections Río Frío 17 and Facatativá 13; Helmens 1990) combined with magnetic polarity dating (Guasca 103; Helmens et al. 1997a). The uplift is thought to have ceased, and the high plain of Bogotá area was situated as presently within the Andean forest belt at about 2,500 m, when the central Bogotá basin started to accumulate sediments at about 3 Ma (lower part of Biozone IV near the base of the Funza II pollen record; Hooghiemstra and Ran 1994; Andriessen et al. 1993; Figs. 2.1 and 2.2). The early part of Biozone IV in the Late Pliocene shows the first possible record of proto-páramo or páramo-like vegetation (Van der Hammen et al. 1973; Helmens and Van der Hammen 1994). Páramo vegetation presently forms the tropic-alpine vegetation belt above the Andean forest limit in the northern Andes.

As a result of the tectonic uplift, the Andes in the Bogotá area possibly lacked sufficiently high enough terrain to support glaciers during the time period preceding the Quaternary. The first mountain glaciation, with glaciofluvial deposition in the Bogotá basin, is recorded near the Gauss/Matuyama magnetic reversal at 2.6 Ma (Helmens et al. 1997a). Since the Funza II pollen record indicates a considerable lowering in the regional forest limit just after ∼2.7 Ma (start of

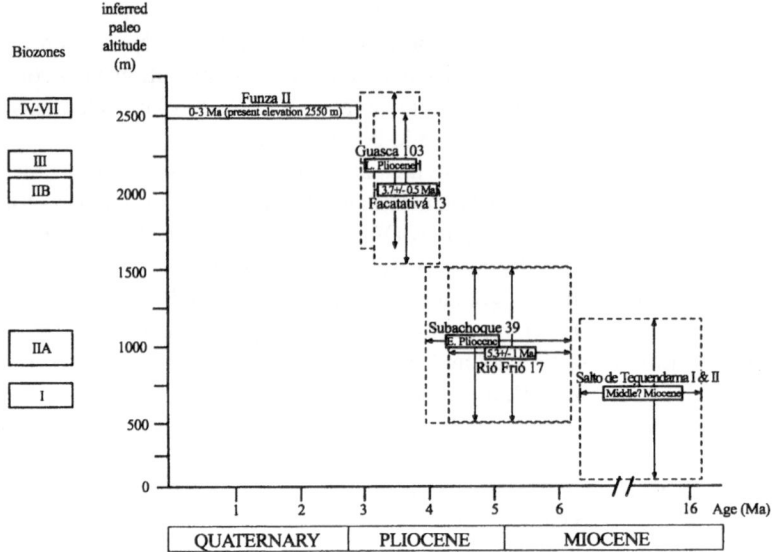

Fig. 2.2 Elevations of past depositional environments, estimated by comparing paleo-floras with present-day equivalents, for a series of sediment sections and the Funza II borehole in the Bogotá area (for locations see Fig. 2.2). Vertical arrows correspond to an estimated uncertainty of ca. 500 m in inferred paleo-altitude. Absolute ages are fission-track dates on intercalated volcanic ashes. Sections make a diagonal in this paleo-altitude versus age diagram, indicating tectonic uplift of the Bogotá region during the Late Miocene and Pliocene prior to ca. 3 Ma. Modified after Wijninga (1996a) and taken from Hooghiemstra et al. (2006)

Biozone IV, upper part; Hooghiemstra and Ran 1994; Andriessen et al. 1993), the first recorded glaciation was most probably related to the global cooling at the base of the Quaternary. An early Matuyama age for the onset of major glaciation in the Bogotá mountains closely corresponds to that obtained for the Bolivian Andes in central South-America, as established by magneto-stratigraphic dating of glacial deposits in the La Paz Basin (Thouveny and Servant 1989).

Major cooling in the Bogotá mountains near the base of the Quaternary is additionally suggested by the occurrence of wide-spread, down-slope movement of old tropical weathering products (Van der Hammen et al. 1973; Helmens 1990). Pollen data obtained from organic intercalations in these slope deposits (San Miguel Formation in Fig. 2.1) show a cold pollen flora corresponding to Biozone IV (upper part) and V (Kuhry and Helmens 1990). The base of the San Miguel Formation has been bracketed by fission-track dates on tephra between 2.8 ± 0.2 and 2.5 ± 0.3 Ma (San Miguel section in Fig. 2.1; P.A.M. Andriessen, K. F. Helmens and R. W. Barendregt, unpublished data).

In the Bogotá area, glacial deposits are restricted to the highest mountain ranges bordering the Bogotá basin, with elevations reaching over 3,600 m. Here a series of moraine complexes dated to the later part of the last glacial cycle, including the Last Glacial Maximum (LGM) at about 20 ka, have been preserved (Helmens et al. 1997b).

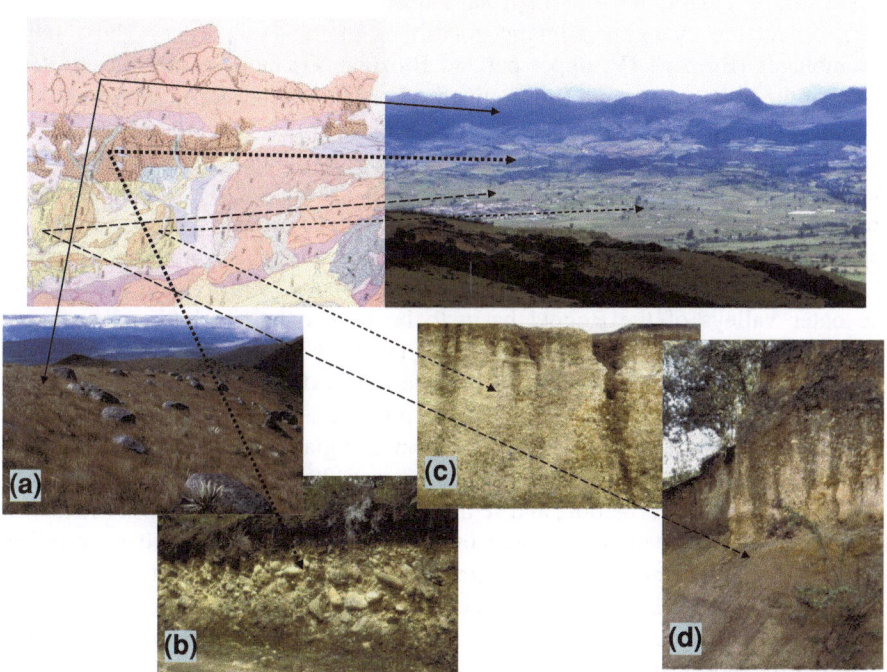

Fig. 2.3 Map fragment (Helmens 1990) and photos illustrating moraines (**a**) and the Río Siecha (**b**: outwash), Río Tunjuelito (**c**: glaciofluvial) and Subachoque Formations (**d**: glaciofluvial alternating with lake sediments) along the western slopes of the Páramo de Palacio and in the adjacent Bogotá basin (Fig. 2.1). Taken from Helmens (2011)

Glaciation During the Early Pleistocene (Matuyama Chron)

In the marginal valleys of the high plain of Bogotá, a distinct, gradual lithological change can be observed from moraine deposits on the higher mountain slopes to glaciofluvial accumulations in the Bogotá basin (Figs. 2.1 and 2.3). The coarse and angular boulders of which the moraines are composed (Fig. 2.3a) pass into more rounded boulders and gravels in a series of large coalescing outwash fans directly at the foot of the formally glaciated mountain slopes (Río Siecha Formation; Figs. 2.1 and 2.3b). The latter deposits grade, within the Bogotá basin, into thick sequences of rounded gravel (Río Tunjuelito Formation; Fig. 2.3c), which in their turn grade into a series of sand and gravel units away from the main rivers that enter the basin. The sand and gravel units alternate with, and in places truncate, more fine-grained sediment beds of mostly lacustrine origin (Subachoque Formation; Figs. 2.1 and 2.3d).

Palynological records from the lacustrine intercalations in the Subachoque Formation, from thin organic beds in the Río Tunjuelito Formation, and from thick sequences of clays interbedded with peaty and sandy sediment in the deeper parts

of the central Bogotá basin (which have been correlated with the type Subachoque), reflect the changing climatic conditions of the Pleistocene. These pollen
assemblages (Biozone IV, upper part, to Biozone VI) indicate that during deposition of the Subachoque and Río Tunjuelito Formations the slopes surrounding the
Bogotá basin were alternately covered by Andean forest vegetation, representing
interglacial (or interstadial) conditions, and treeless páramo vegetation that indicate glacial (stadial) conditions. The sand and gravel interbedded in the Subachoque Formation, and the gravels of the Tunjuelito Formation, are interpreted to
represent the coldest intervals of the Pleistocene when the surrounding mountains
were glaciated (Van der Hammen et al. 1973; Helmens 1990). The glaciers caused
the outer valleys of the Bogotá basin to be infilled by glaciofluvial sediment,
restricting the Bogotá Lake to the central part of the basin. Radiocarbon dates from
the Tunjuelito Formation and from organic-rich sediments and paleosols found
associated with the moraines in the Bogotá mountains suggest synchrony between
the deposition of gravels in the Bogotá basin and glacial events in the mountains
for the Late Pleistocene (Van der Hammen et al. 1980/81; Van der Hammen 1986).

The Subachoque Formation, and sediments of the underlying Guasca Member of
the Upper Tilatá Formation, have been lithologically described in detail and provided
with fission-track dates on volcanic zircons and geomagnetic polarity dates in two
major outcrops, i.e. the Guasca and Subachoque sections along the eastern and western
margins of the Bogotá basin (Fig. 2.1). The clays and silts of the Guasca Member
represent the oldest sediments in the marginal valleys of the Bogotá basin (Helmens
1990). The simplified lithology and absolute chronology of the Guasca and Subachoque sections (Helmens et al. 1997a) are given in Fig. 2.4. The polarity records
obtained from these sediments was not expected to mimic the global geomagnetic
polarity reference timescale considering that the alternating glaciofluvial and lacus-
trine/paludal sediments of the Subachoque and Upper Tilatá Formations accumulated
at different rates, and periods of sedimentation alternated with periods of erosion. The
paleomagnetic correlation made by Helmens et al. (1997a) takes into account the
discontinuity of sedimentation, as well as the error limits of the fission-track dates.

The sudden influx of sands and gravels in the marginal parts of the Bogotá basin
recorded at the base of the Subachoque Formation (Fig. 2.4) is interpreted as
reflecting the onset of glaciation in the adjacent Bogotá mountains (Helmens 1990;
Helmens and Van der Hammen 1994). The sudden influx of more coarse-grained
sediment is accompanied by an increase in the magnitude of the magnetic
susceptibility (MS) signal. Helmens et al. (1997a) use MS as a proxy for periglacial and glacial erosion in the Bogotá mountains, which resulted in a high influx
of magnetite-rich sediments into the basin. The major lithological change at the
base of the Subachoque Formation is bracketed by fission-track dates of
2.9 ± 0.4 Ma and 2.5 ± 0.3 Ma and according to geomagnetic polarity is dated
near the Guass-Matuyama polarity reversal at 2.6 Ma. Glaciations are recorded
throughout the Matuyama Chron. However, the low resolution and fragmentary
nature of the glacial record provided by the Subachoque sediments, and the limited
chronological control, hamper reconstruction of the total number of glaciations in
the Bogotá mountains during the Matuyama as well as detailed land-sea

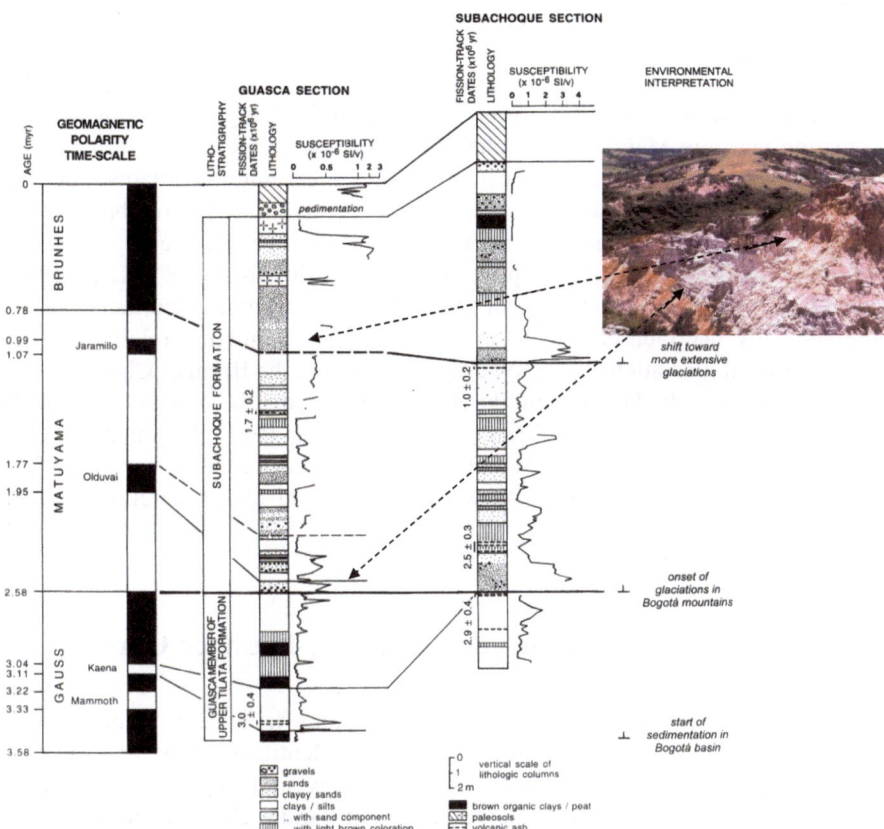

Fig. 2.4 The Late Pliocene-Quaternary environmental record of the Guasca Member of the Upper Tilatá Formation and the Subachoque Formation in the marginal valleys of the Bogotá basin, based on magnetostratigraphy, fission-track chronology, lithology and magnetic suscep-tibility of the Guasca and Subachoque sections (Fig. 2.1). The geomagnetic reversal chronology is based on Cande and Kent (1995). Vertical scale of lithological columns is linear. The sand and gravel interbeds in the Subachoque Formation represent glaciofluvial sediment derived from glaciers in the higher mountain ranges surrounding the Bogotá basin (Helmens et al. 1997a). The photo shows the Guasca section. Taken from Helmens (2011)

correlation. Distinct peaks in the MS sequence suggest at least five glaciations occurred during the Matuyama Chron.

Glaciation During the Middle-Late Pleistocene (Brunhes Chron)

A change in lithology in the upper part of the Subachoque Formation, accompanied by a further increase in the MS signal, is dated at <1.0 ± 0.2 Ma and placed near the Matuyama/Brunhes polarity reversal at 0.8 Ma (Fig. 2.4). Sand and gravel units in the upper part of the Subachoque Formation dated to the Brunhes Chron are distinctly more coarse-grained than those dated to the Matuyama Chron; additionally, peaty horizons and paleosols are found interbedded with the sands and gravels of Brunhes age. Helmens et al. (1997a) interpret the lithological change in the Subachoque Formation near the Matuyama/Brunhes boundary as representing a shift towards more extensive glaciations, which caused rapidly aggrading floodplains to leave a distinct series of coarse-grained sand and gravel layers in the Bogotá basin. Additionally, episodes with conditions warmer than during the Matuyama Chron, and higher evaporation and evapotranspiration rates (Hooghiemstra 1984; Kuhry 1991), probably resulted in lower lake levels, and peat accumulation, and, in the orographically dry Guasca valley, in periods of soil formation (Helmens et al. 1997a). Several glaciations seem to be recorded during the Brunhes Chron.

A shift towards higher magnitude climate oscillations is also recorded in the Funza pollen sequence at ca. 0.8 Ma (Andriessen et al. 1993; Hooghiemstra et al. 1993; Hooghiemstra and Ran 1994). This change has been found associated with a distinct change in the frequency of oscillations from 41 to 100 ka climate cycles (Hooghiemstra et al. 1993).

The Late Pleistocene (Last Glacial Cycle: Brunhes Chron)

Detailed mapping of glacial landforms in the Páramos de Palacio, Sumapáz, Peña Negra and Guerrero (Fig. 2.1) has allowed the identification of four moraine complexes with distinct differences in morphology and degree of denudation (Helmens 1988). Dating of the moraines, and of still older glacial deposits without moraine morphology, using ^{14}C dating on basal lake sediments, peaty horizons and organic-rich paleosols, have dated glacial advances between about 43 and 38 ka BP (in radiocarbon years), 36–31 ka BP, 23.5–19.5 ka BP, 18.0–15.5 ka BP, and 13.5–12.5 ka BP (Helmens 1988; Helmens et al. 1997b).

The most extensive glaciations occurred during the Late Pleistocene in the Middle Wisconsinan probably under the influence of cool and humid conditions (Van der Hammen et al. 1980; Van der Hammen 1981; Helmens and Kuhry 1995). The moraines dated to 18.0–15.5 ka BP (late LGM) show the most impressive morainic morphology of the different morainic complexes recognized. The arcuate, multiple ridge system rises tens of meters above the valley floors and the related maximum ice extent can be continuously traced throughout the mountain ranges studied (Helmens 1988; Fig. 2.1).

Glaciers reached some 100 m further down valley during the early part of the LGM (23.5–19.5 ka BP). The early LGM moraines have been used by Mark and Helmens (2005) to reconstruct paleo-glacier surfaces and equilibrium line altitudes using the area-altitude balance ratio (AABR) method. An overall lowering in ELA from modern values to early LGM of ca. 1,300 m was reconstructed, indicating considerable LGM cooling in the Bogotá mountains. Mark and Helmens (2005), however, do report a large amount of intra-regional variance in LGM ELA that is ascribed to topography and its indirect effect on precipitation, cloudiness and/or glacier form, with lower headwall elevations being correlated to larger accumulation area and lower ELAs. The lowering in ELA of ca. 1,300 m is of similar magnitude to the pollen-based inferred lowering in forest limit in the Bogotá area, implying a drop in mean annual temperature during the early LGM by some 8 °C (Van Geel and Van der Hammen 1973; Kuhry 1988; Helmens et al. 1996).

References

Andriessen PAM, Helmens KF, Hooghiemstra H, Riezebos PA, Van der Hammen T (1993) Absolute chronology of the Pliocene-Quaternary sediment sequence of the Bogotá area, Colombia. Quatern Sci Rev 12:483–501

Cande SC, Kent DV (1995) Revised calibration of the geomagnetic polarity time-scale for the Late Cretaceous and Cenozoic. J Geophys Res 100(B4):6093–6095

Helmens KF (1988) Late Pleistocene glacial sequence in the area of the high plain of Bogotá (Eastern Cordillera, Colombia). Palaeogeogr Palaeoclimatol Palaeoecol 67:263–283

Helmens KF (1990) Neogene-Quaternary Geology in the high plain of Bogotá, eastern Cordillera, Colombia (stratigraphy, paleoenvironments and landscape evolution). Dissertas Botanicae 163, J Cramer, Berlin, p 202

Helmens KF (2004) The Quaternary glacial record of the Colombian Andes. In: Ehlers J, Gibbard PL (eds), Quaternary Glaciations-Extent and Chronology, Part III: South America, Asia, Africa, Australasia, Antarctica. Developments in Quaternary Science, vol. 2c. Elsevier, Amsterdam, pp 115–134

Helmens KF (2011) Quaternary Glaciations of Colombia. In: Ehlers J, Gibbard PL, Hughes PD (eds), Quaternary Glaciations-Extent and Chronology, A Closer Look. Developments in Quaternary Science, vol 15. Elsevier, Amsterdam, pp 815–834

Helmens KF, Kuhry P (1995) Glacier fluctuations and vegetation change associated with Late Quaternary climatic oscillations in the area of Bogotá, Colombia. Quat South America Antarctic Peninsula 9:117–140

Helmens KF, Van der Hammen T (1994) The Pliocene and Quaternary of the high plain of Bogotá (Colombia): a history of tectonic uplift, basin development, and climate change. Quatern Int 21:41–61

Helmens KF, Kuhry P, Rutter NW, Van der Borg K, De Jong AFM (1996) Warming at 18,000 yr BP in the tropical Andes. Quatern Res 48:289–299

Helmens KF, Barendregt RW, Enkin RJ, Bakker J, Andriessen PAM (1997a) Magnetic polarity and fission-track chronology of a Late Pliocene-Pleistocene paleoclimatic proxy record in the Tropical Andes. Quatern Res 48:15–28

Helmens KF, Rutter NW, Kuhry P (1997b) Glacier fluctuations in the eastern Andes of Colombia (South America) during the past 45,000 radiocarbon years. Quatern Int 38–9:39–48

Hooghiemstra H (1984) Vegetational and climatic history of the High Plain of Bogotá, Colombia: a continuous record of the last 3.5 million years. Dissertas Botanicae, vol.79. J Cramer, Vaduz, p 368

Hooghiemstra H, Ran TH (1994) Late Pliocene-Pleistocene high resolution pollen sequence of Colombia: an overview of climate change. Quatern Int 21:63–80

Hooghiemstra H, Melice JL, Berger A, Shackleton NJ (1993) Frequency spectra and paleoclimatic variability of high-resolution 30–1450 ka Funza I pollen record (eastern Cordillera, Colombia). Quatern Sci Rev 12:141–156

Hooghiemstra H, Wijninga VM, Cleef AM (2006) The paleobotanical record of Colombia: implications for biogeography and biodiversity. Ann Missouri Bot Gard 93:297–324

Kuhry P (1988) Palaeobotanical-palaeoecological studies of tropical high Andean peatbog sections (Cordillera Oriental, Colombia). Dissertas Botanicae, vol 116. J Cramer, Berlin

Kuhry P (1991) Comparative hydrogeology in the Andes of Colombia. XIII INQUA Congress, Beijing, China, abstracts, p 179

Kuhry P, Helmens KF (1990) Neogene-Quaternary biostratigraphy and paleoenvironments. In: Helmens KF (ed) Neogene quaternary geology of the high plain of Bogotá, Eastern Cordillera, Colombia (Stratigraphy, Paleoenvironments, and Landscape Evolution), Dissertas Botanicae, vol 163. Cramer, Berlin

Mark BG, Helmens KF (2005) Reconstruction of glacier equilibrium-line altitudes for the last glacier maximum on the high plain of Bogotá, eastern Cordillera, Colombia: climatic and topographic implications. J Quat Sci 20:789–800

Thouret J-C, Van der Hammen T, Salomons B (1996) Paleoenvironmental changes and stades of the last 50,000 years in the Cordillera Central, Colombia. Quatern Res 46:1–18

Thouveny N, Servant M (1989) Paleomagnetic stratigraphy of Pliocene continental deposits of the Bolivian altiplano. Palaeogeogr Palaeoclimatol Palaeoecol 70:331–344

Van der Hammen T (1981) Glaciales y glaciares en el Cuaternario de Colombia: paleoecología y estratigrafía. Rev CIAF 6:635–638 (Bogotá)

Van der Hammen T (1986) La Sabana de Bogotá y su lago en el Pleniglacial Medio. Caldasia 15:249–262

Van der Hammen T, Hooghiemstra H (1997) Chronostratigraphy and correlation of the Pleistocene and Quaternary of Colombia. Quatern Int 40:81–91

Van der Hammen T, Werner JH, Van Dommelen H (1973) Palynology record of the upheaval of the northern Andes: a study of the Pliocene and lower Quaternary of the Colombian eastern Cordillera and the early evolution of its high-Andean biota. Rev Palaeobot Palynol 16:1–122

Van der Hammen T, Duenas H, Thouret JC (1980) Guía de excursión-Sabana de Bogotá. In: Primer Seminario sobre el Cuaternario de Colombia, Bogotá, Colombia. Centro Interamericano de Fotointerpretación (CIAF), Bogotá, 22–29 Aug 1980

Van der Hammen T, Barelds J, De Jong H, De Veer AA (1980/81). Glacial sequence and environmental history in the Sierra Nevada del Cocuy (Colombia). Palaeogeography, Palaeoclimatology, Palaeoecology 32:247–340

Van Geel B, Van der Hammen T (1973) Upper Quaternary vegetational and climatic sequence of the Fúquene area (Eastern Cordillera, Colombia). Palaeogeography Palaeoclimatology Palaeoecology 14:9–92

Wijninga VM (1996a) Paleobotany and palynology of Neogene sediments from the high plain of Bogotá (Colombia). Evolution of the Andean flora from a paleoecological perspective. Unpublished Ph.D thesis, University of Amsterdam, p 370

Wijninga VM (1996b) Neogene ecology of the Salto de Tequendama site (2475 m altitude, Cordillera Oriental, Colombia): the paleobotanical record of montane and lowland forests. Rev Paleobotany Palynology 92:97–156

Wijninga VM (1996c) Palynology and paleobotany of the Early Pliocene section Río Frio 17 (Cordillera Oriental, Colombia): biostratigraphical and chronostratigraphical implications. Rev Paleobotany Palynology 92:329–350

Wijninga VM (1996d) A Pliocene Podocarpus forest mire from the area of the high plain of Bogotá (Cordillera Oriental, Colombia). Rev Paleobotany Palynology 92:157–173

Wijninga VM, Kuhry P (1990) A Pliocene flora from the Subachoque valley (Cordillera Oriental, Colombia). Rev Palaeobot Palynol 62:249–290

Wijninga VM, Kuhry P (1993) Late Pleistocene paleoecology of the Guasca valley (Cordillera Oriental, Colombia). Rev Palaeobot Palynol 78:69–127

Chapter 3
The Glacial and Loess Record of North America

Abstract The best and oldest records of former glaciations are found in the mountainous areas of northwest Canada (Cordillera region) and Alaska, and the Arctic Islands and the northern Interior Plains. Evidence consists of multiple tills, loess units, stratified glacial deposits, nonglacial deposits, and paleosols. Paleomagnetic signals and radiometric dates although scarce, have aided in chronological control. Both mountain and continental glaciations have been recognized and are believed to be roughly synchronous. Two pre-Pleistocene, at least six Early Pleistocene, three Middle Pleistocene, and one Late Pleistocene glaciations have been recognized.

Introduction

In North America, glaciers originated in the extensive and relatively high elevations of the Cordillera (used here to include the various mountain ranges within the Cordilleran region) that commonly reach over 4,000 m and extend from about lat. 65° N in Alaska and Canada to lat. 35° N in the south in the continental United States (Figs. 1.1, 3.1, 3.2). The presence or absence of past and present glaciers in mountainous regions is controlled by variations of snowline elevations becoming lower toward the colder north. Variations of wet Pacific moisture, decreasing generally eastward nourish the glaciers. Tectonism causing uplift, as well as volcanism near the Pacific coast has also affected the presence or absence of glaciers. In addition to glaciers and ice caps, continental ice sheets have covered large parts of Canada and the northern United States in the past. Ice has built up, and coalesced from several centers, controlled by variations of atmospheric circulation and moisture from the Arctic, Pacific, and Atlantic Oceans. Mountains in the Arctic Islands and eastern Canada have also supported glaciers. During at least one

N. Rutter et al., *Glaciations in North and South America from the Miocene to the Last Glacial Maximum*, SpringerBriefs in Earth System Sciences, DOI: 10.1007/978-94-007-4399-1_3, © The Author(s) 2012

glaciation, Cordilleran and continental ice have coalesced. Generally, loess deposits are relatively thin or lacking in much of the glaciated parts of North America. In the unglaciated parts, such as in Alaska and Yukon considerable thicknesses have been found. However, it is in the central part of the United States, outside glacial limits, where thick sequences of interbedded loess units and paleosols are present. Loess units reach over 20 m thick and have commonly been successfully correlated to glacial events (Bobrowsky and Rutter 1992; Levson and Rutter 1996).

Mountain and Continental Glaciation

Evidence for the oldest glaciations in North America is found in the mountainous regions of northwest Canada and Alaska, and the Arctic Islands and the northern Interior Plains (Figs. 3.1, 3.2). Some of the most complete sequences consisting of multiple tills, loess units, stratified glacial deposits, and nonglacial deposits and paleosols are found within the mountain valleys. A good illustration of a multiple till section separated by paleosols is seen in Fig. 3.3 in the Little Bear Creek section, in the Mackenzie Mountains (Fig. 3.1). Correlation of equivalent units is hampered by few absolute dates, and lack of continuous and equivalent sections. Where volcanic rocks and tephra are present, and associated with glacial deposits, age estimates have been determined by K–Ar, ^{40}Ar–^{39}Ar and fission track methods and have aided in establishing a glacial chronology. The youngest deposits have been dated by radiocarbon methods. By far the most important dating method utilized is paleomagnetism. Although a relative dating method, there are enough glacial sequences, with suitable material, spanning thousands to millions of years to determine normal and reversal magnetic trends that aid in interpreting the most likely Chron or Subchron when a certain glaciation took place.

Bostock (1966) was the first to note multiple glaciations in Yukon by recognizing four successively less extensive glaciations—the Early Pleistocene Nansen and Klaza, the Middle Pleistocene Reid, and the Late Pleistocene McConnell. Later workers, in Yukon and adjacent areas added to, and refined the original stratigraphy and concluded that the Nansen and Klaza glaciations were pre-Illinoian (Middle Pleistocene?), the Reid, Illinoian, and the McConnell, Late Wisconsinan (Vernon and Hughes 1966; Rampton 1969). Subsequent work (Duk-Rodkin et al. 1996, 2001; Duk-Rodkin and Barendregt 1997, 2011; Barendregt 2011; Barendregt and Duk-Rodkin 2011) has shown that the Nansen and Klaza glaciations are not recognized outside the reference area and the Reid Glaciation may have taken place before the Illinoian, based upon the age of overlying Sheep Creek tephra (Berger et al. 1996; Westgate et al. 2001; Fig. 3.1).

In northwest North America, evidence for mountain glaciations may go back to over 5 Ma. Excellent sequences are found in the Canadian Cordillera, including the Mackenzie Mountains where there is evidence for more than five glacial events (Duk-Rodkin et al. 1996), in the Ogilvie Mountains at least seven glacial events

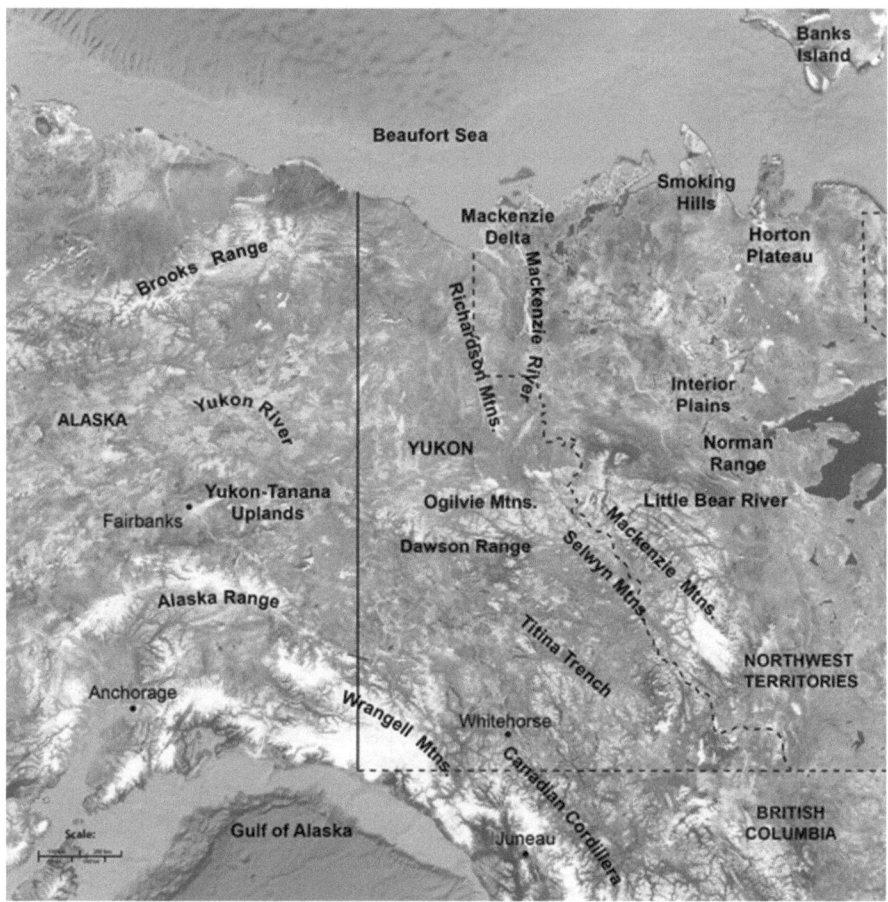

Fig. 3.1 Satellite map of northwest North America (eastern Alaska, Yukon, western northwest Territories) indicating geographical locations and names used in the text

(Duk-Rodkin et al. 2004) and in the Tintina Trench, probably providing the most complete sequence of glaciations, where there is evidence for ten glacial events (Barendregt et al. 2010; Duk-Rodkin et al. 2010; Fig. 3.1). The oldest is of Late Pliocene age, six are from the Early Pleistocene, whereas the next two are Middle Pleistocene, and the youngest Late Pleistocene. In east-central Alaska, Weber (1986) has recognized six glacial events along the Yukon-Tanana Upland, ranging from the Late Tertiary (?) to the Holocene (Fig. 3.1).

Limits of early continental glaciations are difficult to determine in the interior part of northern Canada because the Late Pleistocene ice sheet (LGM) was commonly the most extensive, and eroded much of the evidence for earlier ice. There are some sections in the north that give clues to the distribution of early ice, but generally evidence is scarce. Banks Island is one of the few places where

Fig. 3.2 Satellite map of North America with geographic locations and names used in the text. The *white arrow* points to the general area where many of the best sections indicating multiple glaciation in North America. The *black arrow* points to the general area of widespread loess deposits in the central plains of the United States

shield erratics, moraine landforms, and stratigraphy indicate multiple glaciations (Fig. 3.1; Vincent 1983). On the northern mainland, Pleistocene glacial limits have been estimated from stratigraphical evidence, buried meltwater channels modified

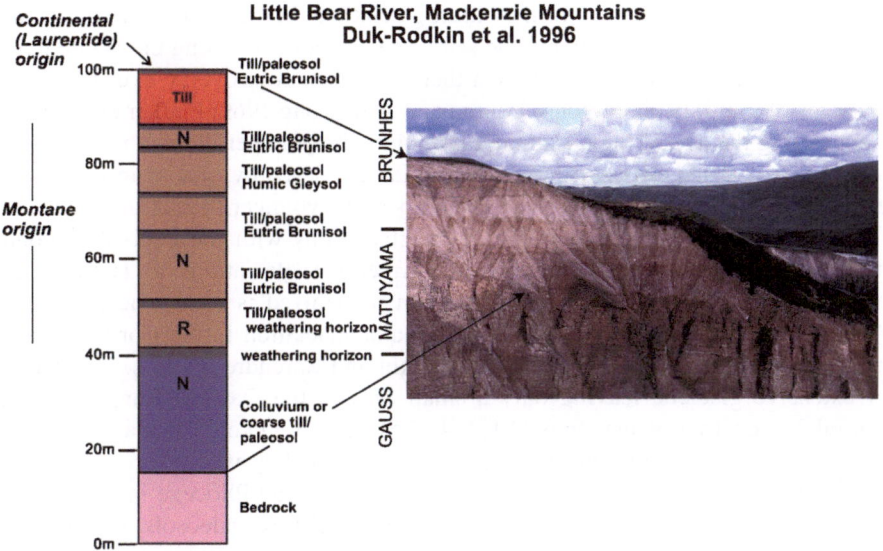

Fig. 3.3 An example of a multiple till and paleosol section from the Mackenzie Mountains, northwest territories. Here, at the Little Bear River section (Fig. 3.1), Continental till (*red*) overlies at least five cordilleran till units (*brown*) alternating with paleosols (*gray*). Paleomagnetic signals are indicated (*N* normal, *R* reversed; after Duk-Rodkin et al. 1996)

by later glaciations, reconstruction of pre-glacial geomorphology to determine ice centers, and use of a climatic/tectonic models to account for distribution of ice (Barendregt and Irving 1998). It appears that ice would have centered on the Horton Plateau during Early Pleistocene glaciations and further south in the Keewatin region during later glaciations (Fig. 3.1). Probably the best evidence for the timing and number of continental glaciers is outside the limits of LGM ice, in the north-central United States (Fig. 3.2). Quaternary sequences consist of multiple till sequences, paleosols and tephras. Paleomagnetic measurements, till composition and characteristics, and tephra dates indicate that ice sheets extended this far south over 2 Ma ago (Roy et al. 2004; Balco and Rovey 2010).

Glaciations During the Late Pliocene: Pleistocene (Gilbert and Gauss/Matuyama, and Matuyama Chrons)

Evidence for the oldest glaciation in northwest North America is found in the glaciomarine Yakataga Formation underlying the Gulf of Alaska that contains ice-rafted debris (Fig. 3.1). Age estimates by Rea and Snoeckx (1995) places the onset of glaciation between 4.2 and about 3.5 Ma. The debris in the glaciomarine Yagataga Formation originated from alpine glaciers in the Alaska coastal ranges.

This indicates that evidence for the oldest glaciations in the interior of Alaska and northwest Canada are most likely the same age or younger. Along the White River in the Wrangell Mountains of Alaska there has been controversy over the origin and ages of multiple diamictons (Denton and Armstrong 1969; Plafker et al. 1977). Considering the age of the Yakataga Formation, the younger diamictons along the White River, dated between 8.9 and 2.7 Ma, are probably glacial in origin and equivalent to the Yakataga Formation and perhaps younger glaciations.

In this study, we combine Cordilleran glaciations with the equivalent continental glaciations and call them North American glaciations to simplify the nomenclature. Therefore, the first glaciation recognized is called the first North American Glaciation. In Yukon, the first major glaciation in the Cordillera took place between 2.90 and 2.58 Ma (Duk-Rodkin and Barendregt 1997). It consisted of extensive glaciers, leaving only a small area of the Dawson Range in west-central Yukon free of ice (Figs. 3.1, 3.4, 3.5). In the Yukon-Tanana Upland the oldest glaciation is found in the Goodpaster River Valley where records of three glaciations are found, with the oldest believed to be Late Pliocene in age. This is based on the interpretation of weathering characteristics, paleosols, stratigraphy, but mainly on magnetostratigraphy (Gauss Chron; Weber 1986). In the Tintina Trench, in central Yukon, two basal units of an extensive section, consists of till overlying pre-glacial gravels, which yielded a Gauss Chron signature (2.90–2.58 Ma). In summary, following the Pliocene glaciation represented by the Yagataga Formation, the next glaciation in northwest North America appears to have taken place in the Late Pliocene (latest Gauss) through perhaps the Early Pleistocene (earliest Matuyama). Although evidence is fragmented, it appears to have been the most widespread glaciation in at least parts of the Cordillera. It is difficult to say the number of advances that took place during this time interval, only to say that it was a period of major ice accumulation in the mountains. Evidence for old continental glaciations is not widespread in the interior part of the northwestern Arctic.

There is questionable evidence for glacial strata near the Smoking Hills in the northern plains that spans the Early Pleistocene Matuyama Chron and possibly the late Gauss Chron (Duk-Rodkin et al. 2004). This would indicate that continental glaciers reached as far as the Mackenzie Delta area from accumulation on the Horton Plateau (first North American Glaciation; Figs. 3.1, 3.4, 3.5).

In the Mackenzie and Ogilvie Mountains there are numerous sections with multiple mountain glacial sequences (Fig. 3.1). The oldest glaciations evidenced by till units, are Early Pleistocene in age (early Matuyama Chron: 1.95–2.58 Ma) dated mainly by paleomagnetism. The oldest is called the Inlin Brook Glaciation, renamed here the second North American Glaciation, and the younger Abraham Glaciation, renamed here the third North American Glaciation (Duk-Rodkin et al. 2004).The long stratigraphic sequence in the Titina Trench near the Yukon/Alaska boundary also yields evidence for several Early Pleistocene glaciations. Cordilleran ice merged from several mountain ranges along the Tintina Trench displaying an outstanding sequence of tills, outwash, and loess revealing six glacial events based on magnetostratigraphy, regional correlation, paleosol

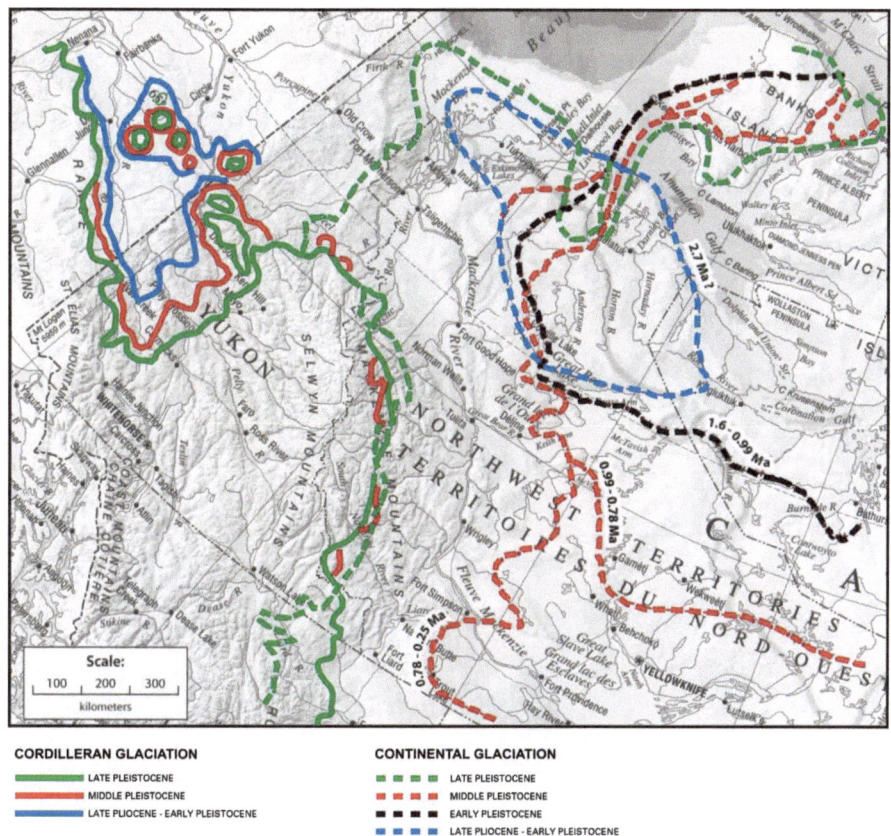

CORDILLERAN GLACIATION

━━━ LATE PLEISTOCENE
━━━ MIDDLE PLEISTOCENE
━━━ LATE PLIOCENE - EARLY PLEISTOCENE

CONTINENTAL GLACIATION

▪ ▪ ▪ ▪ LATE PLEISTOCENE
▪ ▪ ▪ ▪ MIDDLE PLEISTOCENE
▪ ▪ ▪ ▪ EARLY PLEISTOCENE
▪ ▪ ▪ ▪ LATE PLIOCENE - EARLY PLEISTOCENE

Fig. 3.4 Cordilleran/montane and continental ice extent during various time intervals in the late Cenozoic. Positions are approximate. Although incomplete, northwest North America provides the greatest amount of evidence for multiple glaciation in North America (modified from Duk-Rodkin et al. 2004)

development, and pollen content (Barendregt et al. 2010; Duk-Rodkin et al. 2001, 2010). The polarity sequence on the two basal units, till and preglacial gravels, is normal (Gauss Chron: 2.58–2.90 Ma; first North American Glaciation). This is followed by six till/paleosol/loess units of Early Pleistocene age (Matuyama Chron: 2.58–0.78 Ma) consisting of two early advances, the second and third North American Glaciations, an advance during the Olduvai, named here the fourth North American Glaciation, then three post Olduvai advances, the latter during the Jaramillo, named here the fifth North American, sixth North American and seventh North American Glaciations, then by loess/outwash deposited during the final Matuyama glacial event (0.99–0.78 Ma), named the eighth North American Glaciation that extended into an interglacial that spans the Brunhes/Matuyama boundary. These deposits underlie till believed to be Middle

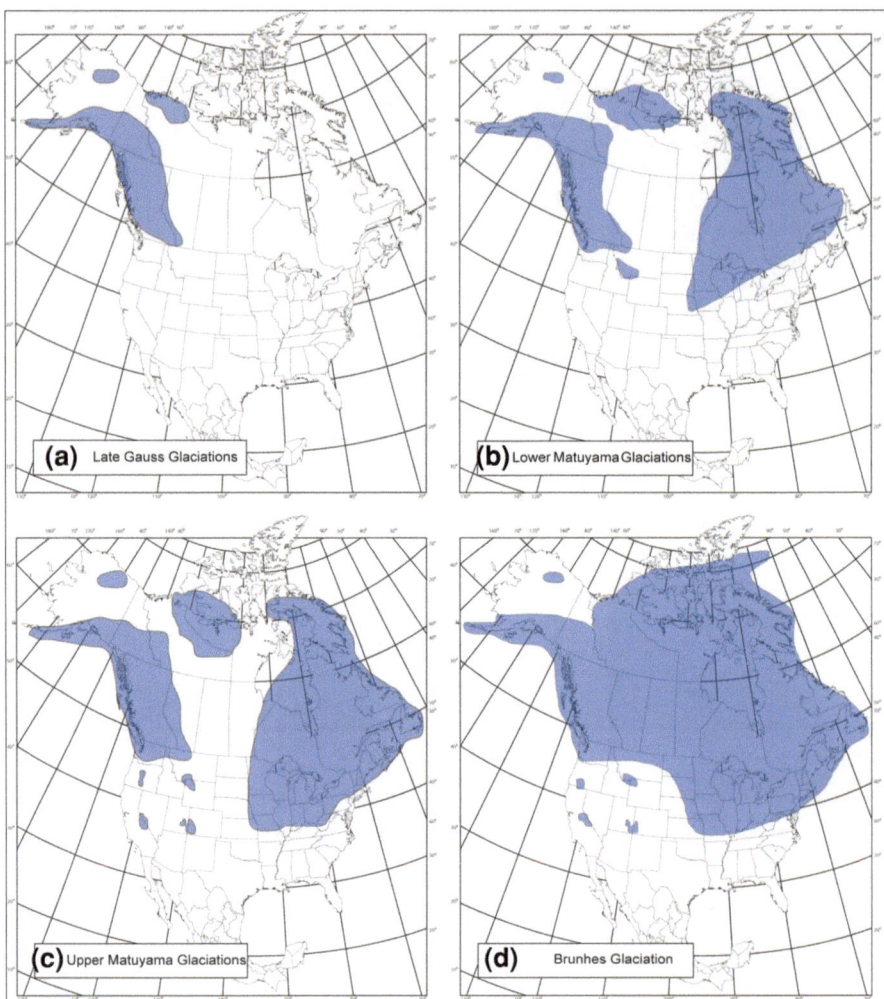

Fig. 3.5 Approximate maximum ice distribution in North America during various time intervals: **a** late Gauss glaciations, **b** lower Matuyama glaciations, **c** upper Matuyama glaciations, and **d** Brunhes glaciations. Modified from Barendregt and Duk-Rodkin (2004)

Pleistocene Reid Glaciation. In the Fort Selkirk area, along the Yukon River, three Early Pleistocene (pre-Reid) glaciations and non-glacial events have been recognized. Deposits lie below lava and hyaloclastic complexes (Jackson et al. 2001) and have been dated by K–Ar, ^{40}Ar–^{39}Ar, fission track, and paleomagnetism. They span 1.83–0.78 Ma within the Matuyama, but from additional data from nearby areas can be further constrained to a glacial event from 1.77 to ca. 1.54 Ma; and another between 1.37 and 0.78 Ma. These are post Olduvai Subchron advances that occurred in the Late Matuyama. Summarizing the above discussion, it is apparent that there have been at least seven Early Pleistocene Cordilleran advances. In some cases, the paleomagnetic data is supplemented with absolute dates allowing more accurate dating of advances.

As mentioned above, evidence for early continental glaciation is scarce in the mainland of northern Canada (Fig. 3.1). The relatively high elevation of the Norman Range—Horton Plateau (1,600 m) would have been the most likely place for continental glaciers to have developed during the Matuyama Chron. Based on Banks Island data of moraines, various glacial and non-glacial sediments and shield erratics, and evidence from the Smoking Hills and the Mackenzie Delta, there were at least five full continental glaciations of various ages that took place. There are as many as four continental advances that took place during the interval 1.77–1.07 Ma. These sediments are covered by interglacial deposits which in turn are overlain by till of Late Matuyama age (<0.99 Ma; Duk-Rodkin et al. 2004). In the Mackenzie Delta, a glacial diamicton was recovered from borehole sediments that were deposited during the interval between 1.25 and 2.25 Ma, thus an Early Matuyama age, and is equivalent to at least one of the tills on Banks Island. The only other place in North America where we can be fairly confident of the ages of Early Pleistocene continental glaciations is in the north-central United States (Fig. 3.2). As mentioned above, using a number or of till criteria, Roy et al. (2004) were able to identify at least four continental glaciations, two of which took place in the Early Matuyama and two that took place in the Late Matuyama. Balco and Rovey (2010) using ^{26}Al-^{10}Be burial isochron methods suggest that continental glaciation in the northern United States took place once in the Early Matuyama, and one in the Late Matuyama. In summary, it is most likely that there were equivalent continental glaciations during most of the major Cordilleran Glaciations. These may have been restricted to small ice caps or more widespread. As can be seen on Figs. 3.4 and 3.5, younger continental glaciations appear to be more widespread than earlier ones, culminating during the Brunhes Chron where the 100 ka cycles dominate as opposed to the earlier 40 ka cycles. It may be, however, that evidence for greater extent of earlier continental glaciations has not yet been detected.

Glaciations During the Middle Pleistocene (Brunhes Chron)

There is evidence for Middle Pleistocene glaciation throughout the mountainous regions of east-central Alaska and Yukon (Figs. 3.1, 3.4, 3.5). The age is based upon degree of preservation of glacial features, the presence of glacial erratics, and its intermediate position between older and younger glacial deposits (Duk-Rodkin et al. 2004).

In Alaska, the Reid Glaciation in northern Canada has been correlated with the Mount Harper Glaciation (Weber 1986), Black Hills Glaciation (Fernald 1965) and Delta Glaciation (Péwé 1975). The age has been determined by a fission track date of 190 ± 20 ka obtained on the Sheep Creek Tephra and a ^{40}Ar–^{39}Ar date of 311 ± 30 ka on a basalt underlying the Reid drift (Duk-Rodkin et al. 2004).

In the Mackenzie Mountains there is a sequence of five mountain tills and one continental till, each unit capped by paleosols (see Fig. 3.3). The three upper units, from the oldest to the youngest, are named the Rouge Mountain (ninth North American Glaciation), Little Keele (tenth North American Glaciation), and the Loreta (Reid equivalent of central Yukon). Although the age of the two mountain tills below the upper mountain till is based upon indirect evidence, they are considered to be Middle Pleistocene (Duk-Rodkin et al. 2004). In the Tintina Trench, outwash and loess overlying Late Matuyama outwash and loess, and underlying Upper Brunhes till and a paleosol is believed to be Middle Pleistocene (Duk-Rodkin et al. 2010). In the mountains of eastern and southern Canada, and the western United States there are no glaciations that can definitely be dated as Middle Pleistocene although there is glacial drift underlying Late Pleistocene deposits in many areas but nothing that can be definitely identified as Middle Pleistocene.

Evidence for Middle Pleistocene continental glaciation is extrapolated from Banks Island to the mainland. The stratigraphy of Banks Island indicates three glaciations each represented by glacial deposits, dated between the Brunhes/Matuyama Chron boundary of 0.78 and 0.13 Ma (Vincent 1983; Barendregt and Irving 1998). These overlie interglacial deposits spanning the Brunhes/Matuyama boundary and underlie glaciomarine sediments dated in less than 100,000 ka, which are considered to be Sangamonan in age (80–130 ka). Other evidence for Brunhes continental glaciation is found in many parts of North America but the best evidence for multiple glaciations is in the north-central United States where three zones of till, each with distinctive properties fall within the Brunhes Chron (Roy et al. 2004). The upper group is underlain by the Lava Creek B ash, dated at 0.602 Ma, leaving the lower two Brunhes tills as Middle Pleistocene glaciations. These are tentatively correlated to what are believed to be Middle Pleistocene glaciations found on Banks Island. The evidence indicates that Middle Pleistocene Cordilleran and continental glaciations were widespread in North America (Fig. 3.4).

Glaciation During the Late Pleistocene (Brunhes Chron)

The Late Pleistocene glaciation (80–10 ka) is by far the best known of all the major glaciations that took place took place in North America. The last advance of the Late Pleistocene left well developed morainic systems and erosional surfaces that are found throughout the major mountain ranges and valleys, and interior parts of Alaska, Canada and northern United States (Figs. 3.1, 3.2, 3.4, 3.5). In North America, the Late Pleistocene is named the Wisconsinan Glaciation following the Sangamon Interglacial. Wisconsinan glaciers expanded around 75 ka years ago, then retreated about 40 ka ago and then re-advanced about 30 ka years ago, where glaciers expanded the furthest in most regions, such as parts of western Canada where it reached its all-time western limit during the latest advance at about 20 ka (Last Glacial Maximum; LGM).

The moisture source of the Cordilleran glaciers was largely from the Pacific Ocean, whereas continental ice, and glaciers that developed in the higher regions of the north and east most likely received moisture from the Arctic and North Atlantic oceans. In the mountains and valleys of northwest North America, the most extensive and latest event is called the McConnell Glaciation, and can be traced in many valleys in Yukon and Alaska (Bostock 1966; Vernon and Hughes 1966; Hughes et al. 1969; Rampton 1969). In the Yukon-Tanana region of Alaska, the McConnell Glaciation is correlated to the Salcha Glaciation (Weber 1986) and in the Mackenzie Mountains the nearest equivalent is the Gayna River Glaciation (Duk-Rodkin and Hughes 1992). There are a number of radiocarbon and ^{36}Cl dates from this region that give an indication of the timing of this last major event (Duk-Rodkin et al. 2004). Expansion of the glaciers probably began by 29 ka ago. The maximum extent of most glaciers was probably somewhere between 24 and 20 ka, and retreated significantly by 9 ka. In the southern Cordillera of western Canada, and in the United States, evidence for the last, most extensive glaciers, is widespread. However, dating control is scarce, but most agree that the Last Glacial Maximum reached its maximum about 20 ka ago (see Ehlers and Gibbard 2007) and had retreated well up the mountain valleys by 11 ka. Late Pleistocene glaciers expanded in the coastal mountains of Alaska and reached their maximum extent about 23 ka ago (Mann and Pettet 1994) and were retreating by or before about 16 ka ago. In contrast, to the south, on northern Vancouver Island, the Last Glacial Maximum took place after ~ 16 ka ago (Al-Suwaidi et al. 2006) and on the Queen Charlotte Islands (Haida Gwaii) glaciers reached their maximum prior to about 16 ka ago, and were retreating by about 15 ka ago (Clague et al. 2004). Late Pleistocene ice terminated in the Puget Sound area of Washington State at about 17 ka ago (Porter and Swanson 1998). As suggested above, in the coastal regions, glacier expansion and retreat took place at different times in different places, probably the result of a more varied climate and perhaps tectonic behavior, in contrast to the eastern Cordillera where expansion of glaciers, during the Last Glacial Maximum, took place before 20 ka ago.

Late Pleistocene continental ice covered a large part of the north and interior plains of Canada, and the northern part of the United States. In western Canada, the ice sheet extended to its all-time maximum, extending within kilometers of the Cordilleran mountain front in the south and into the mountain front in the north. However, in the northern United States, continental ice did not reach as far south as earlier glaciations. In northwest Canada, Last Glacial Maximum continental and mountain ice were not synchronous (Duk-Rodkin et al. 2004). Continental ice reached its maximum, as mountain ice was accumulating in the continental divide area of the Mackenzie Mountains. The continental ice-sheet had already begun retreating when mountain ice reached its maximum extent. In a few localities the two ice masses were able to coalesce at about 30 ka ago. The continental ice then retreated and re-advanced to a maximum extent at about 22 ka ago. As retreat took place, the present day drainage of the Mackenzie River formed at about 12.5 ka ago and several glacial lakes formed between about 11.5 and 10.5 ka ago.

The physical factors, besides climate variations (Milankovitch forcing) that may have caused the presence or variations of glacier extent during the Quaternary, include tectonic uplift in the mountains of northwest North America. Uplift in the late Pliocene aided the formation of glaciers leaving the first records of mountain glaciation. The closure of Panama, sending cool water to the North Pacific, may have aided in glacier expansion as Pacific moisture nourished the glaciers to the east. The 41 ka cycles became more apparent in the Matuyama Chron within increased cooling until about the Jaramillo Subchron where conditions appear to be warmer. During the Brunhes Chron, uplift continued in the Coast Ranges allowing Pacific air to move further eastward. This, along with moisture from the Arctic and Atlantic oceans allowed considerable expansion of continental ice. The 100 ka cycles dominate and increased cooling continued.

The Loess Record

Widespread loess deposits are found in the non-glaciated parts of Alaska, Yukon and north-central United States (Figs. 3.1, 3.2). Although loess deposits may go back to 3 Ma in Alaska (Muhs et al. 2003), loess stratigraphy and how it relates to continental glaciation is best understood in the United States mid-continent area, where investigations have been undertaken for years (see Follmer 1996). Important sources of Quaternary loess in the mid-continent area is mostly from bedrock siltstone and silt of Pliocene eolian deposits to the west and north, carried by westerly or northwesterly winds (Muhs et al. 2008). Glaciogenic silts are generally confined close to the river banks such as along the Platte, Missouri, and south along the east side of the Mississippi. Probably the thickest deposits in the world, formed during last glacial, are found in this region, where deposits are over 20 m thick (Bettis et al. 2003). Although Middle and Late Pleistocene loess/paleosol sequences are found in Alaska and Yukon, deposits with the highest resolution that can be related to glaciation are found south of Late Pleistocene continental glacial deposits (James and Des Moines Lobes) in Nebraska and Iowa. Radiocarbon and optical stimulation dates are used to distinguish events. The oldest loess units are pre-Illinoian in age, consisting of three units, each with a superimposed paleosol or a pedocomplex. Above these units is the Loveland Loess of Illinoian age, equivalent to the Illinoian continental glaciation of Middle Pleistocene age. The Sangamon Soil is developed in Loveland Loess representing the last interglacial period (Rutter et al. 2006). Overlying the Sangamon Soil is the Gilman Canyon Formation (loess) containing the Gilman I and Gilman II paleosols, varying in age from about 27 ka to about 44 ka (Middle Wisconsin; Muhs et al. 2008). The Gilman Canyon Formation could represent the first phase glaciation of the Last

Table 3.1 Identified North American glaciations with age estimates and probable marine isotope stage equivalents

Period	Chron	North American glaciation	Age estimated (Ma)	Probable MIS
Lower miocene-early Pliocene	Gilbert	Yagataga	4.20–3.50	Gi 2, 4, 6, 12, 14, 20, 22, 28 to MG6
Lower Pliocene-early Pleistocene	Gauss/ Matuyama	First North American	2.90–2.58	G10 to 100
Early Pleistocene	E. Matuyama	Second North American	2.58–1.95	100?, 76, 78, 82, 96, 98
Early Pleistocene	E. Matuyama	Third North American	2.58–1.95	78
Early Pleistocene	E. Matuyama/ Olduvai	Fourth North American	1.95–1.77	70
Early Pleistocene	L. Matuyama	Fifth North American	1.77–1.07	50, 52, 54, 58
Early Pleistocene	L. Matuyama	Sixth North American	1.30–1.10	34, 36, 38, 40
Early Pleistocene	L. Matuyama/ Jaramillo	Seventh North American	1.07–0.99	30
Early Pleistocene	L.Matuyama	Eighth North American	0.99–0.78	20, 22
Middle Pleistocene	Brunhes	Ninth North American	0.78–0.28	16, 18
Middle Pleistocene	Brunhes	Tenth North American	0.78–0.28	10, 12
Middle Pleistocene	Brunhes	Reid	0.28–0.13	6, 8
Late Pleistocene	Brunhes	Wisconsinan	0.08–0.02	2, 4

Glacial Maximum (about 80 ka) whereas the Gilman paleosols represent a retreat during warming periods. The overlying Peoria Loess, the thickest loess unit in this region, represents the final and most extensive phase of the Last Glacial Maximum (about 23 ka). The modern soil overlies the Peoria Loess.

In summary, the identified North American Glaciations are listed in Table 3.1.

References

Al-Suwaidi M, Ward BC, Wilson MC, Hebda RJ, Nagorsen DW, Marshall D, Ghaleb B, Wigen RJ, Enkin RJ (2006) Late wisconsian port Eliza cave deposits and their implications for human coastal migration, Vancouver Island, Canada. Geoarchaeology 21:307–332

Balco G, Rovey CW (2010) Absolute chronology for major Pleistocene advances of the laurentide ice sheet. Geology 38:795–798

Barendregt R (2011) Magnetostratigraphy of quaternary sections in eastern Alberta, Saskatchewan and Manitoba. In: Elhers J, Gibbard PL, Hughes PD (eds) Quaternary glaciations—extent and chronology. A closer look. Developments in quaternary science, vol 15, 46. Elsevier, Amsterdam, pp 591–600

Barendregt RW, Duk-Rodkin A (2004) Chronology and extent of late Cenozoic ice sheets in North America: a magnetostratigraphic assessment. In: Ehlers J, Gibbard PL (eds) Quaternary glaciations-extent and chronology. Part II: North America. Developments in quaternary science, vol 2. Elsevier, Amsterdam, pp 1–7

Barendregt RW, Duk-Rodkin A (2011) Chronology and extent of late Cenozoic ice sheets in North America: a magnetostratigraphical assessment. In: Elhers J, Gibbard PL, Hughes PD (eds) Quaternary glaciations—extent and chronology. A closer look. Developments in quaternary science, vol 15, 32. Elsevier, Amsterdam, pp 419–436

Barendregt RW, Irving E (1998) Changes in the extent of North American ice sheets during the late Cenozoic. Can J Earth Sci 35:504–509

Barendregt RW, Enkin RJ, Duk-Rodkin A, White JH (2010) Paleomagnetic evidence for multiple late Cenozoic glaciations in the Tintina Trench, west central Yukon, Canada. Can J Earth Sci 47:987–1002

Berger GW, Péwé TL, Westgate JA, Preece S (1996) Age of sheep creek tephra in central Alaska from thermoluminescence dating of bracketing loess. Quatern Res 45:263–270

Bettis EA, Muhs DR, Roberts HM, Wintle AG (2003) Last glacial loess in the conterminous USA. Quatern Sci Rev 22:1907–1946

Bobrowsky P, Rutter NW (1992) The quaternary geologic history of the Canadian Rocky Mountains. Geographie Physique et Quaternaire 46:5–50

Bostock HB (1966) Notes on glaciation in central Yukon territory. Geological survey of Canada, Paper 65–56, p 18

Clague JJ, Mathewes RW, Ager TA (2004) Environments of northwestern North America before the last glacial maximum. In: Madsen DB (ed) Entering America northeast Asia and Beringia before the last glacial maxima. The University of Utah Press, Salt Lake City, pp 63–94

Denton GH, Armstrong RL (1969) Miocene-Pliocene glaciations in southern Alaska. Am J Sci 267:1121–1142

Duk-Rodkin A, Barendregt RW (1997) Gauss and Matuyama glaciations in the Tintina Trench, Dawson area, Yukon territory. Canadian quaternary association, abstracts, montreal, p 22

Duk-Rodkin A, Barendregt R (2011) Stratigraphical record of glacials/interglacials in northwest Canada. In: Elhers J, Gibbard PL, Hughes PD (eds) Quaternary glaciations—extent and chronology. A closer look. Developments in quaternary science, vol 15, 49. Elsevier, Amsterdam, pp 661–698

Duk-Rodkin A, Hughes OL (1992) Pleistocene montane glaciations in the Mackenzie Mountains, northwest Territories. Géog Phys Quatern 46:69–83

Duk-Rodkin A, Barendregt RW, Tornacai C, Philips FM (1996) Late tertiary to late quaternary record in the Mackenzie Mountains, northwest Territories, Canada: stratigraphy, paleosols, paleomagnetism, and chlorine-36. Can J Earth Sci 33:875–895

Duk-Rodkin A, Barendregt RW, White J, Singhroy VH (2001) Geologic evolution of the Yukon river: implications for gold placer. Quatern Int 80:5–31

Duk-Rodkin A, Barendregt RW, Froese DG, Weber F, Enkin R, Smith IR, Zazula GD, Waters P, Klassen R (2004) Timing and extent of Plio-Pleistocene glaciations in northwestern Canada and east central Alaska. In: Ehlers J, Gibbard PL (eds) Quaternary glaciations-extent and chronology, part II. Developments in quaternary science, vol 2b, Elsevier, Amsterdam, pp 313–345

Duk-Rodkin A, Barendregt RW, White JW (2010) An extensive late Cenozoic record of terrestrial record of multiple glaciations preserved in the Tintina Trunch of west-central Yukon: stratigraphy, paleomagnetism, paleosols, and pollen. Can J Earth Sci 47:1003–1028

Elhers J, Gibbard PL (2007) The extent and chronology of Cenozoic global glaciation. Quatern Int 164–165:6–20

Fernald AT (1965) Glaciation in the Nabesna river area, upper Tanana river valley, Alaska. United States geological survey professional Paper, 525-C, C-120-C-123

Follmer LR (1996) Loess studies in central United States: evolution of concepts. Eng Geol 45:287–304

Hughes OL, Campbell RB, Muller JE (1969) Glacial limits and flow patterns, Yukon territory, south of 65° north latitude. Geological survey of Canada paper 68–34, p 9

Jackson LE Jr., Crystal A, Huscroft R, Gotthardt JE, Storer J, Barendregt RW (2001) Field guide, quaternary volcanism, stratigraphy, vertebrate palaeontology, archaeology, and scenic Yukon river tour, fort Selkirk area (NTS 115 I), Yukon territory. Canadian Quaternary Association, Whitehorse, Yukon Territory, 18–19 Aug 2001, p 24

Levson VM, Rutter NW (1996) Evidence of Cordilleran late wisconsinan glaciers in the 'ice-free corridor'. Quatern Int 32:33–51

Mann DH, Peteet DM (1994) Extent and the timing of the last glacial maximum in southwestern Alaska. Quatern Res 42:136–148

Muhs DR, Ager TA, Bettis EA III, McGeehin J, Been JM, Beget JE, Pavich MJ, Stafford TW Jr, Stevens DSP (2003) Stratigraphy and palaeoclimatic significance of late quaternary loess-palaeosol sequences of the last interglacial-glacial cycle in central Alaska. Quatern Sci Rev 22:1947–1986

Muhs DR, Bettis EA III, Aleinikoff JN, McGeehin JP, Beann J, Skipp G, Marshall BD, Roberts HM, Johnson WC, Benton R (2008) Origin and paleoclimatic significance of late quaternary loess in Nebraska: evidence from stratigraphy, chronology, sedimentology, and geochemistry. Geol Soc Am Bull 120:1378–1407

Péwé TL (1975) Quaternary stratigraphic nomenclature in central Alaska. United States geological survey professional paper 862, p 32

Plafker G, Richter DH, Hudson T (1977) Reinterpretation of the origin of inferred Tertiary tillite in the northern Wrangell Mountains, Alaska. United States geological survey circular 751-B

Porter SC, Swanson TW (1998) Radiocarbon age constraints on rates of advance and retreat of the puget lobe of the cordilleran ice sheet during the last glaciation. Quatern Res 50:205–213

Rampton VN (1969) Pleistocene geology of the Snag-Klutlan area, southwest Yukon, Canada. Unpublished Ph.D dissertation, University of Minnesota, p 279

Rea DK, Snoeckx H (1995) Sediment fluxes in the Gulf of Alaska, the paleoceanographic record from ODP Site 887 on the Patton-Murray seamount platform. In: Rea DK, Basov IA, Scholl DW, Allan JF (eds) Proceedings of the ocean drilling program, pp 247–256

Roy M, Clark PU, Barendregt RW, Glasmann JR, Enkin RJ (2004) Glacial stratigraphy and paleomagnetism of the late Cenozoic deposits of the north-central United States. Geol Soc Am Bull 116:30–41

Rutter NW, Velichko AA, Dlussky KG, Morozova TD, Little EC, Nechaev VP, Evans ME (2006) New insights on the loess/paleosol quaternary stratigraphy from key sections in the U.S. midwest. Catena 67:15–34

Vernon P, Hughes OL (1966) Surficial geology, Dawson, Larsen Creek, and Nash Creek map-areas, Yukon Territory. Geol Surv Can Bull 136: 25

Vincent J-S (1983) La geologie du Quaternaire et la geomorphologie de l'Ile Banks, Arctique Canadien. Geological Survey of Canada Memoire 405: 118

Weber F (1986) Glacial geology of the Yukon-Tanana upland. In: Hamilton T, Reed K, Thorson RM (eds) Glaciation in Alaska, the geologic record. Alaska Geological Society, Anchorage, pp 79–98

Westgate JA, Preece SJ, Froese DG, Walter RC, Sandhu AS, Schweger CE (2001) Dating early and middle (Reid) Pleistocene glaciations in central Yukon, Canada. Quatern Res 56:288–306

Chapter 4
Comparison of South and North American Glaciations

Abstract Comparison of North and South American glaciations is complicated because of incomplete records, lack of age estimates at high resolutions, the determination of what is synchronous and what is non synchronous, and the influence of local climate influences. The glaciations of the two continents are compared by placing their closest equivalents into cold periods determined by marine oxygen isotope records and then comparing them to the Chinese loess/paleosol record (Chap. 5). The earlier glaciations, from about Late Miocene–Early Pliocene (Marine Isotope Stage, MIS >T2) to Early Pleistocene/Early Matuyama (MIS 76) show less equivalence than later glaciations, from about Early Pleistocene/Early Matuyama/Olduvai (MIS 70) to the Late Pleistocene (MIS 2) where several equivalent glaciations can be recognized.

Comparing or correlating glaciations between North and South America is complicated for several reasons. Probably the most important reason is the incomplete records on both continents. In general, the oldest records are the most likely to have been erased by erosion, weathering, and burial. Another problem is dating glacial events at a high enough resolution so that comparisons are possible. Added to these problems are the questions of what time intervals are required for glaciations to be considered synchronous, what is the explanation for synchronous or non-synchronous glaciations, how has active tectonism in mountainous regions affected glaciation, and what is the role of local or regional climate.

We have compared the glaciations of the two continents by placing the closest equivalents into cold periods determined by the marine oxygen isotope record (Lisiecki and Raymo 2005; Shackleton 1995; Opdyke 1995) and then comparing our results to the well-known Chinese loess/paleosol sequence for further verification (Ding et al. 1993, 1994; Rutter 1992; Rutter et al. 1991a, b, 1996).

N. Rutter et al., *Glaciations in North and South America from the Miocene to the Last Glacial Maximum*, SpringerBriefs in Earth System Sciences, DOI: 10.1007/978-94-007-4399-1_4, © The Author(s) 2012

Glaciations During the Miocene-Pliocene (Gilbert and Gauss Chrons) and Early Pleistocene (Matuyama Chron)

The oldest glaciation of South America has been recorded in Lago Buenos Aires, which may very well consist of several advances (Table 4.1, Fig. 3.5). The age control does not allow us to further subdivide Lago Buenos Aires glaciation into separate events. Our best estimate is that there was widespread glaciation between about 7 and 4.4 Ma during cold periods MIS T2 or older (Fig. 3.5). There is no observed equivalent in North America. The oldest we have recorded in North America is the Yagataga Glaciation and it is believed to be younger than Lago Buenos Aires glaciation. Although there is probably more than one glaciation during our estimated time period of about 4.2–3.5 Ma, our age resolution is too low to identify separate glaciations. All we can say is separate glacial advances could have taken place during cold periods MIS Gl28 to MG6, probably mostly during Gi22, Gi20, Gi12, Gi6, Gi4 and Gi2, the coldest periods.

Reliable evidence is available for three glacial events during the Pliocene and near or during the Pliocene/Pleistocene boundary in the Patagonian Andes but not in North America (Table 4.1, Fig. 3.5). These are the Lago Viedma I (about 3.45–3.35 Ma), which probably took place during cold periods MIS MG6, MG8, Lago Argentino I (about 3.30–3.25 Ma) during the cold period M2, and Lago Viedma II (3.00–2.35 Ma) that appears to have occurred near the Pliocene/Pleistocene boundary sometime during cold periods MIS 92, 98, 100, 104, G6, and G10. Equivalent deposits in the Bogotá area, Colombia, may be equivalent to Lago Viedma II. Sometime after Lago Argentino glaciation and during Lago Viedma II glaciation, the first North American Glaciation (2.90–2.58 Ma) is recorded in the mountains of northwestern North America. This event took place most likely during MIS G10 to 100.

In the Early Pleistocene, near the Matuyama/Gauss boundary, there appears to have been widespread glaciation in North America. The second North American Glaciation (2.58–1.95 Ma) occurred at this time and probably took place during MIS 100 but MIS 76, 78, 82, 96, and 98 cannot be ruled out. Not long after, or perhaps overlapping with the second North American Glaciation, Cerro del Fraile II glaciation (<2.144 to >2.132 Ma) took place in South America most likely during MIS 82. Another Early Pleistocene, Matuyama age glaciation that appears to have equivalents on both continents are the Cerro del Fraile III Glaciation (<2.132 to >2.130 Ma) in Argentina and the third North American Glaciation (2.58–1.95 Ma) of northern Canada, and the Interior Plains of Canada and the United States. These glacial events most likely occurred during the very cold period MIS 78. The following Early Pleistocene, Matuyama glaciation, Cerro del Fraile IV (<2.13 to >1.99) has no apparent North American equivalent and is believed to have taken place during MIS 74. Following Cerro del Fraile IV, near synchronous glaciations are recorded in the Early Pleistocene Matuyama Chron. These are the fourth North American Glaciation (1.95–1.77 Ma) and Cerro del Fraile V Glaciation (<1.99 to >1.85 Ma) that took place in MIS 70 and/or 72, in

Table 4.1

Period	Chron	North American glaciation	North American estimated age (Ma)	South American glaciation	South American estimated age (Ma)	Probable MIS
Late Miocene-early Pliocene	Gilbert	None		Lago Buenos Aires	7.00–5.00	>T2
Early and late Pliocene	Gilbert	Yagataga	4.20–3.50	None		Gi 2, 4, 6, 12, 14, 20, 22, 28–MG6
Late Pliocene	Gauss	None		Lago Viedma I	3.45–3.35	MG6–MG8
Late Pliocene	Gauss	None		Lago Argentino	3.30–3.25	M2
Late Pliocene-early Pleistocene	Gauss/Matuyama	First North American	2.90–2.58	None		G10–100
Late Pliocene-early Pleistocene	Gauss/Matuyama	None		Lago Viedma II	3.00–2.35	92, 96, 98, 100, 104, G6, G10
Early Pleistocene	Early Matuyama	None		Cerro del Fraile I	2.18 to >2.6	90, 98, 100, 104
Early Pleistocene	Early Matuyama	Second North American	2.58–1.95	None		100, 76, 78, 82, 96, 98
Early Pleistocene	Early Matuyama	None		Cerro del Fraile II	<2.144 to >2.132	78, 82
Early Pleistocene	Early Matuyama	Third North American	2.58–1.95	Cerro del Fraile III	<2.132 to >2.130	78
Early Pleistocene	Early Matuyama	None		Cerro del Fraile IV	<2.130 to >1.99	74, 76
Early Pleistocene	Early Matuyama/Olduvai	Fourth North American	1.95-1.77	Cerro del Fraile V	<1.99 to >1.85	70, 72
Early Pleistocene	Late Matuyama	Fifth North American	1.77–1.07	Cerro del Fraile VI	<1.78 to >1.43	50, 52, 54, 58
Early Pleistocene	Late Matuyama	Sixth North American	1.30–1.10	None		34, 36, 38, 40
Early Pleistocene	Late Matuyama/Jaramillo	Seventh North American	1.07–0.99	Great Patagonian Glaciation	1.15–1.05	30

(continued)

Table 4.1 (continued)

Period	Chron	North American glaciation	North American estimated age (Ma)	South American glaciation	South American estimated age (Ma)	Probable MIS
Early Pleistocene	Late Matuyama	Eigth North American	0.99–0.78	Post GPG I	<1.10 to >0.76	20, 22
Middle Pleistocene	Brunhes	Ninth North American	0.78–0.28	Post GPG II	<0.76	16, 18
Middle Pleistocene	Brunhes	Tenth North American	0.78–0.28	None		10, 12
Middle Pleistocene	Brunhes	Reid	0.28–0.13	Post GPG III	0.26–0.12	6, 8
Late Pleistocene	Brunhes	Wisconsinan	0.08–0.02	Late glacial maximum	0.048–0.025	2, 4

the Olduvai Subchron. Following these events, Early Pleistocene, Matuyama advances on both continents took place most likely sometime during MIS 50, 52, 54 and 58. These are the fifth North American Glaciation (1.77–1.07 Ma) and Cerro del Fraile VI (<1.78 to >1.43 Ma). Following these events, the Early Pleistocene, Matuyama sixth North American Glaciation (1.30–1.10 Ma) took place most likely sometime during cold periods MIS 34 to 40. The Early Pleistocene Great Patagonian Glaciation (1.15–1.05 Ma), the most extensive recognized in South America, took place during the Jaramillo Subchron, during MIS 30. This advance is paired with the seventh North American Glaciation (1.07–0.99 Ma) that most likely took place at about the same time. The last glaciations recorded in the Early Pleistocene, Matuyama Chron, are the Post Great Patagonian Glaciation I (<1.05 to >0.76 Ma) and its probable equivalent in North America, the eigth North American Glaciation (0.99–0.78 Ma), fairly tightly constrained in MIS 22 and 20.

Glaciations During the Middle and Late Pleistocene (Brunhes Chron)

The glaciations of the Middle Pleistocene fall in the extensive period between about 0.780–0.120 Ma (Table 4.1, Fig. 3.5). In the Middle Pleistocene, Brunhes Chron, the first glaciation recorded is the Post Great Patagonian II Glaciation (<0.76 Ma) equated to the ninth North American Glaciation (0.78–0.28 Ma), that most likely took place sometime after MIS 18 or 16. There is no South American equivalent for the following glaciation which is only recorded in North America. This is named the tenth North American Glaciation (0.78–0.28 Ma) and took place sometime during MIS 10 and 12. The last two glaciations recorded appear to have near synchronous equivalents on both continents. These are the Post Great Patagonian III Glaciation (0.26–012 Ma) of South America and the Reid Glaciation recorded the best in Yukon, Canada (0.28–0.13 Ma). They occurred most likely during MIS 6, although there is evidence in Yukon that the Reid occurred during MIS 8.

Finally, the last major glacial cycle, by far the glaciation with the most widespread evidence and dating control on both continents took place during the Late Pleistocene. In North America it is named the Wisconsinan Glaciation (0.08–0.02 Ma), with evidence widely available in the mountainous regions and the interior plains of Canada and the United States. In South America, called the Last Glacial Maximum (0.048–0.025 Ma), evidence is evident in the Patagonian and Colombian Andes. In addition, widespread loess deposits found in the Interior Plains of North America and the Pampean region of Argentina are equated to this glaciation. In North America, there is evidence for glacial advances during MIS 4 and 2, whereas in South America the dating is more elusive, suggesting that the major ice flow was during MIS 2. Table 4.1 and Fig. 4.1 summarizes the comparisons and correlation of glaciations.

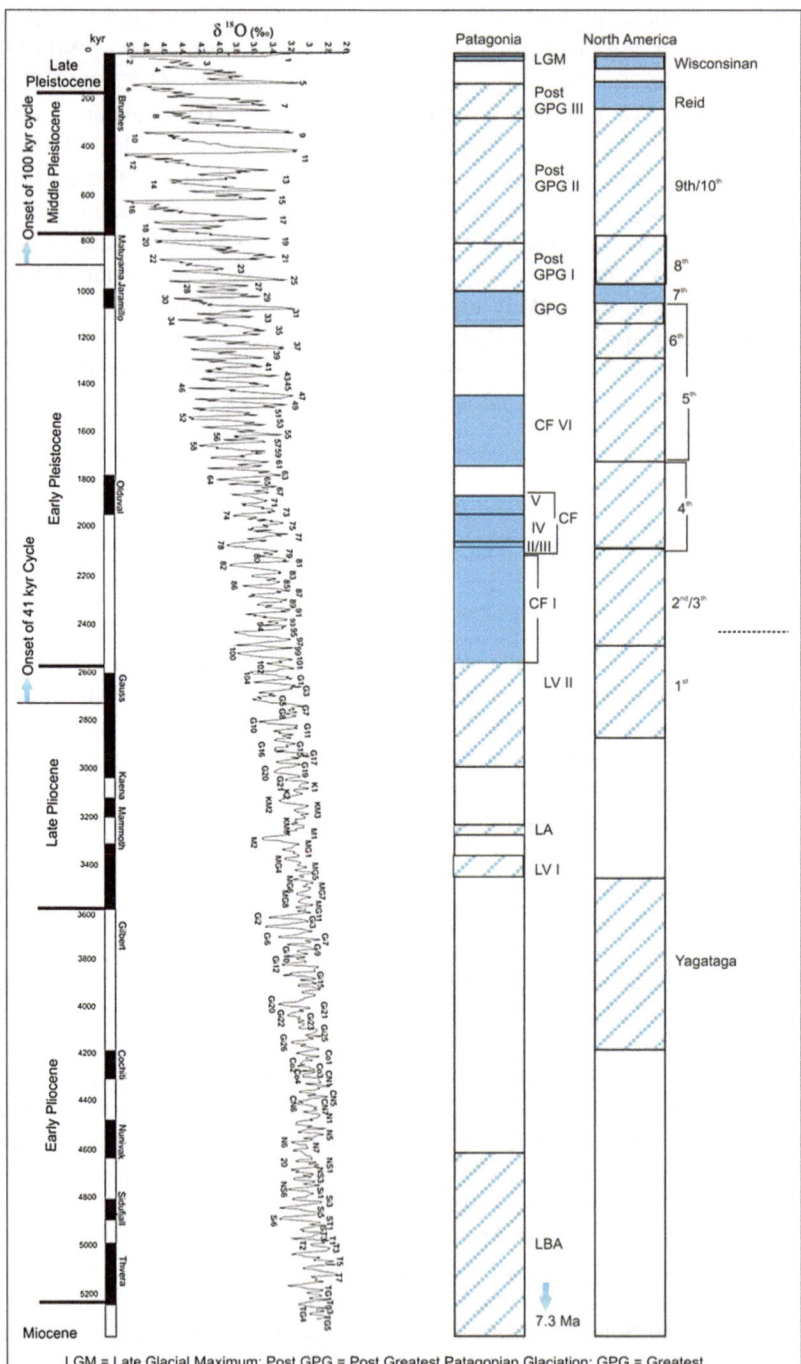

LGM = Late Glacial Maximum; Post GPG = Post Greatest Patagonian Glaciation; GPG = Greatest Patagonian Glaciation; CF = Cerro del Fraile; LV = Lago Viedma; LA = Lago Argentino, LBA = Lago Buenos Aires. Solid bars = well defined boundaries, diagonal dashed bars = poorly defined boundaries. MIS equivalents from Lisiecki and Raymo (2005).

◄ **Fig. 4.1** The LR04 benthic δ^{18}O stack constructed by the graphic correlation of 57 globally distributed benthic δ^{18}O records. The stack is plotted using the LR04 age model, and with new MIS labels for the early Pliocene (see Lisiecki and Raymo 2005). The glaciations identified in these studies are equated to the best fit MIS cold periods. Note the vertical scale changes across panels (after Lisiecki and Raymo 2005)

References

Barendregt RW, Duk-Rodkin A (2004) Chronology and extent of Late cenozoic ice sheets in North America: a magnetostratigraphic assessment. In: Ehlers J, Gibbard PL (eds) Quaternary glaciations-extent and chronology. Part II: North America. Developments in quaternary science vol 2. Elsevier, Amsterdam, pp 1–7

Ding ZL, Rutter NW, Liu TS (1993) Pedostratigraphy of Chinese loess deposits and climatic cycles in the last 2.5 Ma. Catena 20:73–91

Ding ZL, Yu Z, Rutter NW, Liu TS (1994) Towards an orbital time scale for Chinese loess deposits. Quatern Sci Rev 13:39–70

Lisiecki LE, Raymo ME (2005) A Pliocene-Pleistocene stack of 57 globally distributed benthic δ^{18}O records. Paleoceanography 20(PA 1003):1–17

Opdyke ND (1995) Mammalian migration and climate over the last 7 Ma. In: Vrba ES et al (eds) Paleoclimate and evolution, with emphasis on human origins. Yale University Press, New Haven, pp 109–114

Rutter NW (1992) Presidential address, XIII INQUA congress 1991: Chinese loess and global change. Quatern Sci Rev 11:275–281

Rutter NW, Ding Z, Evans ME, Liu TS (1991a) Boaji-type pedostratigraphic section, Loess Plateau, North-Central China. Quatern Sci Rev 10:1–22

Rutter NW, Ding Z, Liu TS (1991b) Comparison of isotope stages 1–61 with the Baoji-type pedostratigraphic section of North-Central China. Can J Earth Sci 28:985–990

Rutter NW, Ding Z, Liu TS (1996) Long paleoclimatic records from China. Geophysica 32:7–34

Shackleton NJ (1995) New data on the evolution of Pliocene climatic variability. In: Vrba ES et al (eds) Paleoclimate and evolution, with emphasis on human origins. Yale University Press, New Haven, pp 242–248

Chapter 5
Chinese Loess/Paleosol Record

Abstract The loess/paleosol sequence of north central China is probably the most complete record of Quaternary climate change derived from terrestrial deposits in the world. Alternating loess units (utilizing grain-size differences) which indicate cold periods and paleosols which indicate warm periods, record 37 major climatic cycles over the past 2.6 Ma. The Eastern Hemisphere loess/paleosol climate records are compared to the Western Hemisphere glaciations in order to record similarities of climate change intervals over a wider area by comparing Marine Isotope Stage (MIS) records. Results indicate that even though there are many uncertainties in timing of glaciations, there are major equivalent synchronous climate intervals represented in both areas.

In order to expand our understanding of past Western Hemisphere glaciation to the Eastern Hemisphere, we compare our results with the well-known Chinese loess paleosol records. Briefly, the Loess Plateau of north central China is underlain by probably the most complete and continuous terrestrial Late Cenozoic record on Earth (Liu 1985; Kukla 1987; Rutter et al. 1996). The Quaternary record consists of alternating loess units with superimposed soils (paleosols), representing over 37 major climate cycles over the last 2.6 Ma (Rutter et al. 1991a, b; Ding et al. 1993). More recent studies show that the Late Pliocene red clay formation below the Quaternary sequence extends the record back to about 7 Ma (Ding et al. 1998, 1999, 2000). Recently, the loess/paleosol sequences have been extended back to the Early Miocene (21.4 Ma; Hao and Guo 2007). It can be difficult to identify climate changes in the red clay formation, because it consists of paleosol complexes that mask distinct loess units. There is enough differentiation, however, to say that major climate changes took place back to at least 21 Ma. Here, we compare or glacial record back to the loess/paleosol records to the beginning of the Quaternary.

The loess is derived from the north, northwest, and west of the Loess Plateau, from the Gobi Desert and perhaps from glacial deposits in the mountains of

N. Rutter et al., *Glaciations in North and South America from the Miocene to the Last Glacial Maximum*, SpringerBriefs in Earth System Sciences, DOI: 10.1007/978-94-007-4399-1_5, © The Author(s) 2012

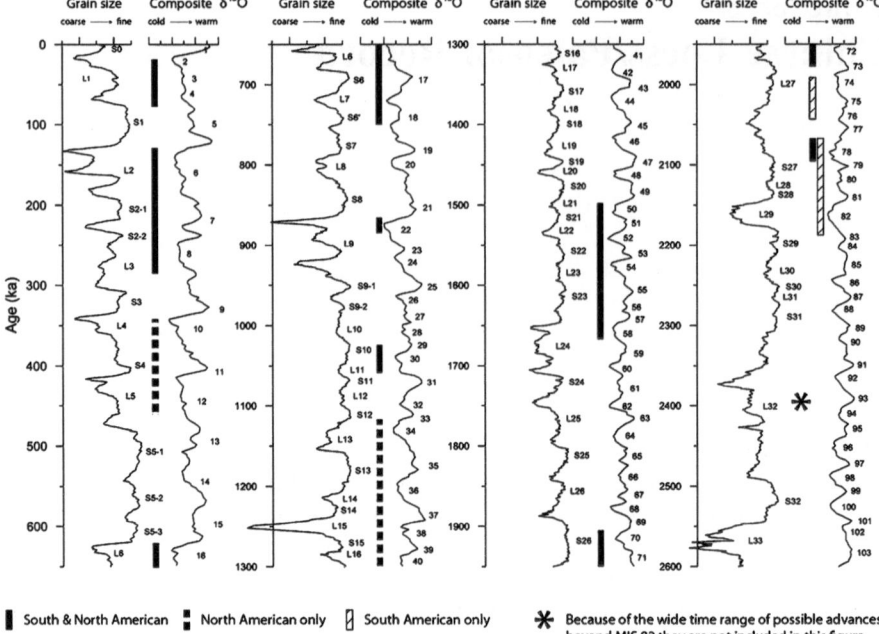

Fig. 5.1 Correlation of grain-size records of Chinese loess with a composite marine oxygen isotope record. The relatively coarse grain-sizes indicate cold periods which are correlated to MIS cold periods (see Ding et al. 2002). The glaciations identified from South and North America are then compared to the MIS and grain-size records. The most probable cold periods when glaciations took place are shown by vertical bars

Afghanistan. An additional source area could be the continental shelf off the west Pacific margin exposed at lower sea levels. The Quaternary deposits consisting mostly of fine sand, silt, and clay are over 400 m in the northwest part of the Plateau, thinning to the southeast to about 200 m. Climate today in the northern part of the Chinese Loess Plateau is arid (<200 mm annual precipitation) becoming more temperate toward the south (>800 mm annual precipitation).

These conditions persisted with some modifications throughout the Quaternary, and probably earlier, at least, from the Late Pliocene. The variations in climate throughout the Quaternary were determined by a number of climate proxies including variations in the texture of the loess, soil type, and magnetic susceptibility. Determining the age of the various units consisted of radiocarbon, thermoluminescence and optical stimulation methods, sedimentation rates, paleomagnetic signatures and an orbital time scale developed from the loess record (Ding et al. 1994, 2001). The textural characteristics of the loess proved to be an important climate proxy (Rutter 1992; Fig. 5.1). The relatively thick, coarse grained loess units were deposited during extensive periods of cold, dry winter monsoon winds driving dust from the north, northwest, and west. The source and dominant wind directions were essentially constant for at least the Quaternary Period. This resulted in overall grain-size decreasing toward the south and southeast and equivalent units becoming thinner. The winter monsoon is controlled largely by the position and intensity of the Siberian high pressure system. The stronger the winds, the further large grains can be carried. The superimposed soils (paleosols) in the loess units provide another important climate proxy. The soils are similar to grassland to mixed forest-grassland soils. The soils were developed when warm, moist summer monsoons from the south dominated for periods similar to that of the winter monsoons. During these periods, the rate of loess deposition and grain-size were greatly reduced, or deposition ceased all together. These long periods allowed soils to develop fully even as some deposition took place.

The alternation of the loess/paleosol units are best explained by the Milankovitch theory of orbital forcing as is the oxygen isotope record used in this study. The relatively complete record of loess deposition and paleosol formation, and the grain-size characteristics of the loess units, controlled largely by the influence of the winter and summer Monsoons, was the key in development of a grain-size orbital time scale (Ding et al. 1994, 2002). Figure 3.5 plots the grain-size orbital time scale against a stacked oxygen isotope record. The same climate variation frequencies in each record can be correlated. Even though the oxygen isotope record utilized here was constructed with different orbital time scale (Ding et al. 2002) than the one used in correlating the North and South America glacial events (Lisiecki and Raymo 2005), they are close enough that the timing of the glacial events are not compromised at the scale we are using. We can therefore, interpret and correlate climate events from the grain-size record with the oxygen isotope records used in our glacial correlation back to about 2.6 Ma. The clear relationships between these records, taken from widely different locations, suggest that major climatic events were worldwide in scope.

Prior to the Early Pleistocene, it is difficult to compare records because of the low resolution of the glacial record and the complicated record of pre-Pleistocene loess/paleosol sequences (Fig. 5.1). The data record major climatic changes but they are not clear enough to correlate at an acceptable level. At about the Pleistocene-Pliocene boundary (Gauss-Matuyama boundary), the grain-size data clearly show extensive cold periods (MIS 100–103) supporting the glacial records of both continents, Lago Viedma II and at least the end of the first and second North American glaciations. This was undoubtedly a time of extensive worldwide glaciation. As revealed by loess grain-size data, the dominant cold period MIS 82 supports Cerro del Fraile II Glaciation and with equivalent N and S glaciations during MIS 78, the third North American and Crerro del Fraile III. Although the cold periods appear weak during MIS 74 to 76, Cerro del Fraile IV probably took place during this time frame. The equivalent glaciations, the fourth North American and Cerro del Fraile V, most likely occurred during MIS 72, the stronger cold period relative to MIS 70. After a period of no recorded glacial activity from about MIS 69 to 59, although the grain-size and MIS records indicate cold periods, the next equivalent glacial records occur between MIS 58 to 50. These are the fifth. North American and Cerro del Fraile VI glaciations. Following this activity, no recorded glaciations are apparent between MIS 49 and 41, although cold periods occurred. The sixth North American Glaciation is recorded between MIS 40 and 34 and most likely occurred during MIS 34, 36 or 38, prominent cold periods. No South American equivalents are recorded. Near synchronous glaciations become more apparent when the dominant 100 ka cycles begin after about 1 Ma. The Great Patagonian and the seventh North American glaciations took place most likely during MIS 30, although the grain-size record shows a weak cold period. MIS 22 indicates a strong cold period in the grain-size record when the post GPG I and eighth North American Glaciation took place and appear to be equivalent. MIS 18 and 16 show strong cold periods in the grain–size record when the post GPG II and the ninth North American Glaciation advanced. After a period of a weak cold period, MIS 14, where no glaciation is recorded, North American glacial activity resumed during MIS 12 and 10, during relatively strong cold periods when the 10th North American Glaciation occurred. Cold periods recorded by the grain-size records resulted in synchronous glaciations in both North and South America during MIS 8, 6, the Post GPG III and Reid, and during MIS 4 and 2, the Last Glacial Maximum and the Wisconsinan.

Comparing the cold periods indicated by the loess grain-size record from Asia with the western hemisphere glacial records shows that the marine record (oxygen isotope ratios from marine fauna) support the grain-size and glacial records, although does not clear up many uncertainties in timing of the glaciations. It can be said that there were major worldwide synchronous climate change intervals. In addition, with the paucity of glacial evidence, and the low resolution and number of the age estimates available, the record of glaciation is far from complete. The questions of what cold periods supported glaciation and what cold periods did not, and why, still have to be answered.

References

Barendregt RW, Duk-Rodkin A (2004) Chronology and extent of late cenozoic ice sheets in North America: a magnetostratigraphic assessment. In: Ehlers J, Gibbard PL (eds) Quaternary glaciations-extent and chronology. Part II: North America, developments in quaternary science, vol 2. Elsevier, Amsterdam, pp 1–7

Ding ZL, Rutter NW, Liu TS (1993) Pedostratigraphy of Chinese loess deposits and climatic cycles in the last 2.5 Ma. Catena 20:73–91

Ding ZL, Yu Z, Rutter NW, Liu TS (1994) Towards an orbital time scale for Chinese loess deposits. Quatern Sci Rev 13:39–70

Ding ZL, Sun JM, Liu TS, Zhu RX, Yang SL, Guo B (1998) Wind-blown origin of the pliocene red clay formation in the Central Loess Plateau, China. Earth Planet Lett 161:135–143

Ding ZL, Xiong SF, Sun JM, Yang SL, Gu ZY, Liu TS (1999) Pedostratigraphy and paleomagnetism of about 7 Ma eolian loess-red clay sequence at Lingtai, Loess Plateau, North-Central China and the implications for paleomonsoon evolution. Palaeogeogr Palaeoclimatol Palaeoecol 152:49–66

Ding ZL, Rutter NW, Sun JM, Yang SL, Liu TS (2000) Re-arrangement of atmospheric circulation at about 2.6 Ma over Northern China: evidence from grain size records of loess-palaesol and red clay sequences. Quatern Sci Rev 19:547–558

Ding ZL, Yang SL, Hou X, Wang Z, Chen Z, Liu TS (2001) Magnetostratigraphy and sedimentology of the Jingchuan red clay section and correlation of the tertiary eolian red clay sediments of the Chinese Loess Plateau. J Geophys Res 106:6399–6407

Ding ZL, Derbyshire E, Yang SL, Yu ZW, Xiong SF, Liu TS (2002) Stacked 2.6 Ma grain size record from the Chinese loess based on five sections and correlation with the deep-sea $\delta^{18}O$ record. Paleoceanography 17:1–21

Hao Q, Guo Z (2007) Magnetostratigraphy of an early–middle miocene loess-soil sequence in the Western Loess Plateau of China. Geophys Res Lett 34:L18305. doi:10.1029/2007GL031162

Kukla G (1987) Loess stratigraphy in Central China. Quatern Sci Rev 6:191–219

Lisiecki LE, Raymo ME (2005) A pliocene–pleistocene stack of 57 globally distributed benthic $\delta^{18}O$ records. Paleoceanography 20(PA1003):1–17

Liu TS (ed) (1985) Loess and the environment. China Ocean Press, Beijing 251

Rutter NW (1992) Presidential address, XIII INQUA congress 1991: Chinese loess and global change. Quatern Sci Rev 11:275–281

Rutter NW, Ding Z, Evans ME, Liu TS (1991a) Boaji-type pedostratigraphic section, Loess Plateau, North-Central China. Quatern Sci Rev 10:1–22

Rutter NW, Ding Z, Liu TS (1991b) Comparison of isotope stages 1–61 with the Baoji-type pedostratigraphic section of North-Central China. Can J Earth Sci 28:985–990

Rutter NW, Ding Z, Liu TS (1996) Long paleoclimatic records from China. Geophysica 32:7–34

Discussion and Conclusions

From our discussions in comparing the glaciations of North and South American glaciations, there is little doubt that the astronomical theory or Milankovich forcing is the overall control of the extensive glaciations recorded in the two hemispheres. This is fortified by comparing our results with the Chinese loess/ paleosol record, suggesting that astronomical forcing has controlled the Earth's major glaciations from at least the Miocene Epoch. Not all glaciations have been recorded, not least of which is due to glacial evidence eroded by natural and artificial processes, undiscovered evidence, and incomplete and poor spatial and age estimates. With the exception of glaciations that are recorded in only one hemisphere, most glaciations are roughly correlated, some more accurately than others. We are aware that our control is commonly poor, and that speculation is a concern. This is manifested by the low resolution of our age estimates and the lack of control of spatial relationships of equivalent glaciations. However, there are many factors that alter the timing and extent of glaciations that are otherwise controlled by Milankovitch forcing. It is well known that climate has become generally cooler from at least the Pliocene Epoch to the present (Lisiecki and Raymo 2005). We also know that the climatic cycles in the Quaternary Period and probably before, have not been constant but have shifted from a dominating periodicity of about 41 ka to a periodicity of 100 ka at about 1 Ma years ago and remains that way today. It is difficult to explain the reason for this shift and may involve, among other things, complex feedback mechanisms, such as interplay of Milankovitch forcing, ocean and atmosphere circulation variations, tectonic activity causing altitudinal changes, and ice sheet build up. For example, prior to 900,000 years ago North American ice sheets may have melted during precessional or obliquity-induced warm stages, but at around 900,000 years, ice sheet build up due to critical temperature cooling allowed ice to persist through weaker insolation maxima and therefore, grow to a larger size over each successive cycle (Raymo 1997). Denton (2000) suggested that ice sheets grow steadily over the course of the 100 ka cycle, extracting waters from the Earth's

N. Rutter et al., *Glaciations in North and South America from the Miocene to the Last Glacial Maximum*, SpringerBriefs in Earth System Sciences, DOI: 10.1007/978-94-007-4399-1, © The Author(s) 2012

oceans and leading to major reorganization of deep-water circulation from an "interglacial" to a "glacier" mode. Eventually collapse takes place, influenced by atmospheric forcing, and the cycle begins again. The role of CO_2 appears to be an important factor in this process (Ruddiman 2003). Shackleton (2000) has demonstrated how at the 100 ka cycle, atmospheric CO_2, air temperature, and deep water temperatures are in phase, whereas ice volume lags these three variables, and therefore the 100 ka cycles enters the paleo-record according to the concentration of atmospheric CO_2. In our comparative study, the dominance of the 100 ka cycles after about 900,000 years increased the reliability of our correlations by providing higher resolution and accuracy in identifying MIS cold periods. The higher frequencies but shorter time periods of cold periods before about 900,000 years hindered our chances of accurate correlations within a single marine isotope cold period because of the low resolution of our age estimates. This commonly resulted in the best correlation within more than one MIS cold period.

Tectonism, along with its influence on weathering and global CO_2 distribution, may have played an important part in the results that we have presented. Most of our glacial evidence comes from the mountains of western South and North America. There is evidence that the Patagonian Andes rose more rapidly in the south than in the north, changing atmospheric circulation, providing rain shadows that generated arid and semiarid conditions in the extra Patagonian Andes, and to some extent in the southern Pampas, enabling loess beds to be deposited. This diachronous uplift probably reduced extension of the glaciers out of the mountains. As uplift continued in the northern Andes, how rain shadows affected the extent of glaciations is not clear. In northwest Canada and Alaska, there is evidence that as the coastal mountains rose during the Late Miocene and continued through the Quaternary Period, there was a gradual shift of snowfall eastward resulting in relatively more widespread glaciation in the eastern Cordillera and the plains to the east. Although the details of tectonism are elusive, we can say that tectonism has had effect in our North–South comparisons and correlations.

Late Miocene and Pliocene glaciations appear to be more abundant or easier to detect in the Patagonian Andes than in northwest North America. A possible explanation for this is the influence of Antarctic cooling that has caused northwest flow of cold water along the coast of South America, adding to the development of Andean glaciers at least since the Late Miocene (Rabassa et al. 2005).

The widespread glaciation that took place in the Northern Hemisphere at the beginning of the Quaternary (Gauss/Matuyama boundary) could be attributed to the closure of the Isthmus of Panama and rerouting of ocean currents as proposed by Bartoli et al. (2005), who suggested the closure took place between 3.1 and 2.8 Ma ago. However, there was widespread glaciation in South America at about the same time.

Other observations that could alter direct correlation of glaciations between the continents include variations of sea surface temperatures through time. Investigations by Becquey and Gersonde (2002) show that Pleistocene summer sea-surface temperatures (SSST), reconstructed from planktonic foraminifera from cores in the Subantarctic-Zone of the South Atlantic, show that cold climatic conditions prevailed at lat. 43°S during both glacial and interglacial periods from about 1.83 to 0.87 Ma. Between about 0.9 and 0.4 Ma higher amplitude fluctuations are observed in the SSST between glacial and interglacial periods corresponding to the temperature range of the present Polar Front and Subantarctic Front. These variations may be related to changes in the Northern Hemisphere ice sheets. The past 0.4 Ma are characterized by strong SSST variations, of up to 8 °C.

There are other factors that could influence our correlations, such as the 'bipolar seesaw' identified in ice-cores, as proposed by Broecker (1998). Correlation of ice-cores from Greenland and Antarctica indicate a global-scale anti-phase relationship, whereby cooling in the Antarctica coincides with warming in Greenland and vice-versa. This was consistent during the last glacial cycle.

In conclusion, we have compared and correlated glaciations between North and South America. Even though the age estimates are at a low resolution and not all glacial evidence is available, it is clear that major glaciations fall within broad time intervals and therefore, can be called roughly synchronous. They can be explained by Milankovitch forcing but modified by a variety of natural external factors that influence the presence, volume, and extent of glaciations both spatially and temporally.

References

Bartoli G, Sarnthein M, Eeinelt H (2005) Final closure of Panama and the onset on northern hemisphere glaciation. Earth Planet Sci Lett 237:33–44

Becquey S, Gersonde R (2002) Past hydrographic and climate changes in the subarctic zone of the south Atlantic-the Pleistocene record from ODP site 1090. Palaeogeogr Palaeoclimatol Palaeoecol 182:221–239

Broecker WS (1998) Paleocean circulation during the last deglaciation: a bipolar seesaw? Paleoceanography 13:119–121

Denton GH (2000) Does an asymmetric thermocline-ice-sheet oscillator drive 100,000 year glacial cycles? J Quat Sci 15:301–318

Lisiecki LE, Raymo ME (2005) A pliocene–pleistocene stack of 57 globally distributed benthic $\delta^{18}O$ records. Paleoceanography 20(PA 1003):1–17

Rabassa J, Coronato AMJ, Salemme M (2005) Chronology of the late Cenozoic patagonian glaciations and their correlation with biostratigraphic units of the Pampean region (Argentina). J S Am Earth Sci 20:81–103

Raymo ME (1997) The timing of major climatic terminations. Paleoceanography 12:577–585

Ruddiman WS (2003) Orbital insolation, ice volume, and greenhouse gases. Quatern Sci Rev 22:1597–1629

Shackleton NJ (2000) The 100,000-year ice-age cycle identified and found to lag temperature, carbon dioxide, and orbital eccentricity. Science 289:1897–1902